Tucholsky Wagner Zola Scott Sydow Freud Schlegel
Turgenev Fonatne Wallace
Twain Walther von der Vogelweide Fouqué Friedrich II. von Preußen
Weber Freiligrath
Fechner Fichte Weiße Rose von Fallersleben Kant Ernst Frey Frommel
Engels Fielding Hölderlin Richthofen
Fehrs Faber Flaubert Eichendorff Tacitus Dumas
Feuerbach Maximilian I. von Habsburg Fock Eliasberg Zweig Ebner Eschenbach
Ewald Eliot Vergil
Goethe Elisabeth von Österreich London
Mendelssohn Balzac Shakespeare Dostojewski Ganghofer
Trackl Lichtenberg Rathenau Doyle Gjellerup
Stevenson Hambruch
Mommsen Tolstoi Lenz Droste-Hülshoff
Thoma Hanrieder
Dach von Arnim Hägele Hauff Humboldt
Reuter Verne
Karrillon Rousseau Hagen Hauptmann Gautier
Garschin
Damaschke Defoe Hebbel Baudelaire
Descartes
Hegel Kussmaul Herder
Wolfram von Eschenbach Dickens Schopenhauer
Bronner Darwin Melville Grimm Jerome Rilke George
Campe Horváth Aristoteles Bebel Proust
Bismarck Vigny Barlach Voltaire Federer Herodot
Gengenbach Heine
Storm Casanova Tersteegen Grillparzer Georgy
Chamberlain Lessing Langbein Gilm Gryphius
Brentano Lafontaine
Strachwitz Claudius Schiller Kralik Iffland Sokrates
Katharina II. von Rußland Bellamy Schilling
Gerstäcker Raabe Gibbon Tschechow
Löns Hesse Hoffmann Gogol Wilde Vulpius
Luther Heym Hofmannsthal Gleim
Roth Klee Hölty Morgenstern Goedicke
Heyse Klopstock Kleist
Luxemburg Puschkin Homer Mörike
La Roche Horaz Musil
Machiavelli Kierkegaard Kraft Kraus
Navarra Aurel Musset
Nestroy Marie de France Lamprecht Kind Kirchhoff Hugo Moltke
Laotse Ipsen Liebknecht
Nietzsche Nansen
Marx Ringelnatz
von Ossietzky Lassalle Gorki Klett Leibniz
May vom Stein Lawrence Irving
Petalozzi Knigge
Platon Pückler Michelangelo Kafka
Sachs Poe Kock
de Sade Praetorius Mistral Liebermann Zetkin Korolenko

The publishing house tredition has created the series **TREDITION CLASSICS**. It contains classical literature works from over two thousand years. Most of these titles have been out of print and off the bookstore shelves for decades.

The book series is intended to preserve the cultural legacy and to promote the timeless works of classical literature. As a reader of a **TREDITION CLASSICS** book, the reader supports the mission to save many of the amazing works of world literature from oblivion.

The symbol of **TREDITION CLASSICS** is Johannes Gutenberg (1400 – 1468), the inventor of movable type printing.

With the series, tredition intends to make thousands of international literature classics available in printed format again – worldwide.

All books are available at book retailers worldwide in paperback and in hardcover. For more information please visit: www.tredition.com

tredition was established in 2006 by Sandra Latusseck and Soenke Schulz. Based in Hamburg, Germany, tredition offers publishing solutions to authors and publishing houses, combined with worldwide distribution of printed and digital book content. tredition is uniquely positioned to enable authors and publishing houses to create books on their own terms and without conventional manufacturing risks.

For more information please visit: www.tredition.com

Practical Mechanics for Boys

James Slough Zerbe

Imprint

This book is part of the TREDITION CLASSICS series.

Author: James Slough Zerbe
Cover design: toepferschumann, Berlin (Germany)

Publisher: tredition GmbH, Hamburg (Germany)
ISBN: 978-3-8491-5384-7

www.tredition.com
www.tredition.de

CONTENTS

Introductory

I. On Tools Generally

Varied Requirements. List of Tools. Swivel Vises. Parts of Lathe. Chisels. Grinding Apparatus. Large Machines. Chucks. Bench Tools. Selecting a Lathe. Combination Square. Micrometers. Protractors. Utilizing Bevel Protractors. Truing Grindstones. Sets of Tools. The Work Bench. The Proper Dimensions. How Arranged.

II. How to Grind and Sharpen Tools

Importance of the Cutting Tool. The Grinder. Correct Use of Grinder. Lathe Bitts. Roughing Tools. The Clearance. The Cutting Angle. Drills. Wrong Grinding. Chisels. Cold Chisels. System in Work. Wrong Use of Tools.

III. Setting and Holding Tools

Lathe Speed. The Hack-saw. Hack-saw Frame. The Blade. Files. Grindstones. Emery and Grinding Wheels. Carelessness in Holding Tools. Calipers. Care in Use of Calipers. Machine Bitts. The Proper Angle for Lathe Tools. Setting the Bitt. The Setting Angle. Bad Practice. Proper Lathe Speeds. Boring Tools on Lathe. The Rake of the Drill. Laps. Using the Lap. Surface Gages. Uses of the Surface Gage.

p. ii

IV. On the First Use of the File

The First Test. Filing an Irregular Block. Filing a Bar Straight. Filing Bar with Parallel Sides. Surfacing Off Disks. True Surfacing. Precision Tools. Test of the Mechanic. Test Suggestions. Use of the Dividers. Cutting a Key-way. Key-way Difficulties. Filing Metal Round. Kinds of Files. Cotter-file. Square. Pinion. Half-round. Round. Triangular. Equalizing. Cross. Slitting. Character of File Tooth. Double Cut. Float-cut. Rasp Cut. Holding the File. Injuring Files. Drawing Back the File.

V. How to Commence Work

Familiarity with Tools. File Practice. Using the Dividers. Finding Centers. Hack-saw Practice. Cutting Metal True. Lathe Work. First Steps. Setting the Tool. Metals Used. The Four Important Things. Turning Up a Cylinder. Turning Grooves. Disks. Lathe Speeds.

VI. Illustrating Some of the Fundamental Devices

Belt Lacing. Gears. Crown Wheel. Grooved Friction Gearing. A Valve which Closes by the Water Pressure. Cone Pulleys. Universal Joint. Trammel for Making Ellipses. Escapements. Simple Device to Prevent a Wheel or Shaft from Turning Back. Racks and Pinions. Mutilated Gears. Simple Shaft Coupling. Clutches. Ball and Socket Joints. Tripping Devices. Anchor Bolt. Lazy Tongs. Disk Shears. Wabble Saw. Crank Motion by a Slotted Yoke. Continuous Feed by Motion of a Lever. Crank Motion. Ratchet Head. p. iii Bench Clamp. Helico-volute Spring. Double helico-volute. Helical Spring. Single Volute Helix Spring. Flat Spiral, or Convolute. Eccentric Rod and Strap. Anti-dead Center for Lathe.

VII. Properties of Materials

Elasticity. Traction. Torsion. Flexure. Tenacity. The Most Tenacious Metal. Ductility. Malleability. Hardness. Alloys. Resistance. Persistence. Conductivity. Equalization. Reciprocity. Molecular Forces. Attraction. Cohesion. Adhesion. Affinity. Porosity. Compressibility. Elasticity. Inertia. Momentum. Weight. Centripetal Force. Centrifugal Force. Capillary Attraction. The Sap of Trees. Sound. Acoustics. Sound Mediums. Vibration. Velocity of Sound. Sound Reflections. Resonance. Echos. Speaking Trumpet. The Stethoscope. The Vitascope. The Phonautograph. The Phonograph. Light. The Corpuscular Theory. Undulatory Theory. Luminous Bodies. Velocity of Light. Reflection. Refraction. Colors. The Spectroscope. The Rainbow. Heat. Expansion.

VIII. How Draughting Becomes a Valuable Aid

Lines in Drawing. Shading. Direction of Shade. Perspectives. The Most Pronounced Lines. Direction of Light. Scale Drawings. Degree, and What it Means. Memorizing Angles. Section Lining. Making

Ellipses and Irregular Curves. Focal Points. Isometric and Perspective. The Protractor. Suggestions in Drawing. Holding the Pen. Inks. Tracing Cloth. Detail Paper. How to Proceed. Indicating Material by Section Lines. p. iv

IX. Treatment and Use of Metals

Annealing. Toughness and Elasticity. The Process. Tempering. Tempering Contrasted with Annealing. Materials Used. Gradual Tempering. Fluxing. Uniting Metals. Alloying Method. Welding. Sweating. Welding Compounds. Oxidation. Soldering. Soft Solder. Hard Solder. Spelter. Soldering Acid. The Soldering Iron.

X. On Gearing, and How Ordered

Spur and Pinion. Measuring a Gear. Pitch. Diametral Pitch. Circular Pitch. How to Order a Gear. Bevel and Miter Gears. Drawing Gears. Sprocket Wheels.

XI. Mechanical Power

The Lever. Wrong Inferences from Use of Lever. The Lever Principle. Powers vs. Distance Traveled. Power vs. Loss of Time. Wrongly-Directed Energy. The Lever and the Pulley. Sources of Power. Water Power. Calculating Fuel Energy. The Pressure or Head. Fuels. Power from Winds. Speed of Wind and Pressure. Varying Degrees of Pressure. Power from Waves and Tides. A Profitable Field.

XII. On Measures

Horse Power. Foot Pounds. Energy. How to Find Out the Power Developed. The Test. Calculations. The Foot Measure. Weight. The Gallon. The Metric System. Basis of Measurement. Metrical Table, Showing Measurements in Feet and Inches. p. v

XIII. Useful Information for the Workshop

Finding the Circumference of a Circle. Diameter of a Circle. Area of a Circle. Area of a Triangle. Surface of a Ball. Solidity of a Sphere. Contents of a Cone. Capacity of a Pipe. Capacity of Tanks. To Toughen Aluminum. Amalgams. Prevent Boiler Scaling. Diamond Test. Making Glue Insoluble in Water. Taking Glaze Out of Grindstone. To Find Speeds of Pulleys. To Find the Diameters Required. To Prevent Belts from Slipping. Removing Boiler Scale. Gold Bronze. Cleaning Rusted Utensils. To Prevent Plaster of Paris from Setting Quickly. The Measurement of Liquids with Spoons.

XIV. Simplicity of Great Inventions and of Nature's Manifestation

Invention Precedes Science. Simplicity in Inventions. The Telegraph. Telephone. Transmitter. Phonograph. Wireless Telegraphy. Printing Telegraph. Electric Motor. Explosions. Vibrations in Nature. Qualities of Sound. The Photographer's Plate. Quadruplex Telegraphy. Electric Harmony. Odors. Odophone. A Bouquet of Vibrations. Taste. Color.

XV. Workshop Recipes and Formulas

Adhesives for Various Uses. Belt Glue. Cements. Transparent Cement. U. S. Government Gum. To Make Different Alloys. Bell-metal. Brass. Bronzes. Boiler Compounds. Celluloid. Clay Mixture for Forges. Modeling Clay. Fluids for Cleaning Clothes, Furniture, etc. Disinfectants. Deodorants. Emery for Lapping Purposes. Explosives. Fulminates. Files, and How to p. vi Keep Clean. Renewing Files. Fire-proof Materials or Substances. Floor Dressings. Stains. Foot Powders. Frost Bites. Glass. To Frost. How to Distinguish. Iron and Steel. To Soften Castings. Lacquers. For Aluminum and Brass. Copper. Lubricants. Paper. Photography. Plasters. Plating, Coloring Metals. Polishes. Putty. Rust Preventives. Solders. Soldering Fluxes. Steel Tempering. Varnishes. Sealing Wax.

XVI. Handy Tables

Table of Weights for Round and Square Steel. Table of Weight of Flat Steel Bars. Avoirdupois Weight. Troy Weight. Apothecaries' Weight. Linear Measure. Long Measure. Square Measure. Solid or Cubic Measure. Dry Measure. Liquid Measure. Paper Measure. Table of Temperatures. Strength of Various Metals. Freezing Mixtures. Ignition Temperatures. Power and Heat Equivalents.

<table><tr><td>XVII.</td><td>Inventions and Patents, and Information About the Rights and Duties of Inventors and Workmen</td></tr></table>

The Machinist's Opportunities. What is an Inventor? Idea Not Invention. What an Invention Must Have. Obligation of the Model Builder. Paying for Developing Devices. Time for Filing an Application. Selling an Unpatented Invention. Joint Inventors. Joint Owners Not Partners. Partnerships in Patents. Form of Protection Issued by the Government. Life of a Patent. Interference Proceedings. Concurrent Applications. Granting Interference. Steps in Interference. First Sketches. First Model. First Operative Machine. p. vii Preliminary Statements. Proving Invention. What Patents Are Issued For. Owner's Rights. Divided and Undivided Patents. Assignments. How Made. What an Invention Must Have. Basis for Granting Patent in the United States. Reasons for Granting Abroad. Original Grants of Patents. International Agreement. Application for Patents. Course of Procedure. Costs. Filing a Matter of Secrecy.

Glossary of Words

LIST OF ILLUSTRATIONS

FIG.

1. Bench vise
2. Pipe grip for vise
3. Swivel vise
4. Speed lathe
5. Calipers
6. Engine lathe
7. Center gage
8. Pocket screw and wire gage
9. Handy bench vise
10. Combination square
11. Uses of the combination square
12. A quick adjusting micrometer
13. Universal bevel protractor
14. Uses of universal bevel protractor
15. Grindstone truing device
16. Set of tools and case
17. The work bench
18. Hook tool
19. Parting tool
20. Knife tool
21. Right-hand side tool
22. Internal tool
23. Left-hand side tool
24. Tool for wrought iron
25. Tool for cast iron
26. End view of drill
27. Side view of drill
28. Hack-saw frame
29. Hack-saw blade
30. Plain hook tool
31. Plain straight tool
32. Proper angles for tools

33. Angles for tools
34. Angles for tools
35. Set of the bitt
36. Correct angle
37. Wrong angle
38. Too low
39. Improper set
40. Internal set
41. Set for brass
42. Surface gage
43. Uses of surface gage
44. Rounded surface
45. Winding surface
46. Hexagon nut
47. Laying off hexagon nut
48. Cutting key-way
49. Key-seat rule
50. Filing metal round
51. Filing metal round
52. Making a round bearing
53. Making a round bearing
54. Cross section of file
55. Files
56. Correct file movement
57. Incorrect file movement
58. Belt lacing
59. Belt lacing
60. Belt lacing
61. Belt lacing
62. Bevel gears
63. Miter gears
64. Crown wheel
65. Grooved friction gears
66. Valve

67. Cone pulleys
68. Universal joint
69. Trammel
70. Escapement
71. Device for holding wheel
72. Rack and pinion
73. Mutilated gears
74. Shaft coupling
75. Clutches
76. Ball and socket joints
77. Fastening ball
78. Tripping devices
79. Anchor bolt
80. Lazy tongs
81. Disc shears
82. Wabble saw
83. Continuous crank motion
84. Continues feed
85. Crank motion
86. Ratchet head
87. Bench clamp
88. Helico-volute spring
89. Double helico-volute
90. Helical spring
91. Single volute-helix
92. Flat spiral or convolute
93. Eccentric rod or strap
94. Anti dead-centers for lathes
95. Plain circle
96. Ring
97. Raised surface
98. Sphere
99. Depressed surface
100. Concave

101. Forms of cubical outlines
102. Forms of cubical outlines
103. Forms of cubical outlines
104. Forms of cubical outlines
105. Shading edges
106. Shading edges
107. Illustrating heavy lines
108. Illustrating heavy lines
109. Lines on plain surfaces
110. Lines on plain surfaces
111. Illustrating degrees
112. Section lining
113. Drawing an ellipse
114. Perspective at angles
115. Perspective of cube
116. Perspective of cube
117. Perspective of cube
118. Protractor
119. Using the protractor
120. Section-lining metals
121. Spur gears
122. Miter gear pitch
123. Bevel gears
124. Laying of miter gears
125. Sprocket wheel
126. Simple lever
127. Lever action
128. The pulley
129. Change of direction
130. Change of direction
131. Steam pressure
132. Water pressure
133. Prony brake
134. Speed indicator

p. 1

PRACTICAL MECHANICS FOR BOYS

INTRODUCTORY ToC

The American method of teaching the mechanical arts has some disadvantages, as compared with the apprentice system followed in England, and very largely on the continent.

It is too often the case that here a boy or a young man begins work in a machine shop, not for the avowed purpose of learning the trade, but simply as a helper, with no other object in view than to get his weekly wages.

Abroad, the plan is one which, for various reasons, could not be tolerated here. There he is bound for a certain term of years, and with the prime object of teaching him to become an artisan. More often than otherwise he pays for this privilege, and he knows it is incumbent on him "to make good" right from the start.

He labors under the disadvantage, however, that he has a certain tenure, and in that course he is not pushed forward from one step to the next on account of any merit of his own. His advancement is fixed by the time he has put in at each p. 2 part of the work, and thus no note is taken of his individuality.

Here the boy rises step after step by virtue of his own qualifications, and we recognize that one boy has the capacity to learn faster than another. If he can learn in one year what it requires three in another to acquire, in order to do it as perfectly, it is an injury to the apt workman to be held back and deterred from making his way upwardly.

It may be urged that the apprentice system instills thoroughness. This may be true; but it also does another thing: It makes the man a mere machine. The true workman is a thinker. He is ever on the alert to find easier, quicker and more efficient means for doing certain work.

What is called "Efficiency" in labor methods, can never obtain in an apprenticeship system for this reason. In a certain operation, where twelve motions are required to do a certain thing, and a minute to perform the twelve operations, a simplified way, necessitating only eight motions, means a difference in saving one-third of the time. The nineteen hundred fewer particular movements in a day's work, being a less strain on the operator, both physically and mentally, to say nothing whatever of the advantages which the proprietor of the shop would gain. p. 3

I make this a leading text in the presentation of this book; namely, that individual merit and stimulus is something of such extreme importance that it should be made the keynote for every boy who tries to become a mechanic.

The machinist easily occupies a leading place in the multitude of trades and occupations. There is hardly an article of use but comes to the market through his hands. His labor is most diverse, and in his employment doing machine work he is called upon to do things which vary widely in their character.

These require special knowledge, particular tools, and more frequently than otherwise, a high order of inventive ability to enable him to accomplish the task.

The boy should be taught, at the outset, that certain things must be learned thoroughly, and that habits in a machine shop can be bad as well as good. When he once becomes accustomed to putting a tool back in its rightful place the moment he is through with it, he has taken a long step toward efficiency.

When he grasps a tool and presents it to the work without turning it over several times, or has acquired the knack of picking up the right tool at the proper place, he is making strides in p. 4 the direction of becoming a rapid and skilled workman.

These, and many other things of like import, will require our attention throughout the various chapters.

It is not the intention of the book to make every boy who reads and studies it, a machinist; nor have we any desire to present a lot of useful articles as samples of what to make. The object is to show the boy what are the requirements necessary to make him a machinist; how to hold, handle, sharpen and grind the various tools; the proper ones to use for each particular character of work; how the various machines are handled and cared for; the best materials to use; and suggest the numerous things which can be done in a shop which will pave the way for making his work pleasant as well as profitable.

It also analyzes the manner in which the job is laid out; how to set the tools to get the most effective work; and explains what is meant by making a finished piece of workmanship. These things, properly acquired, each must determine in his own mind whether he is adapted to follow up the work.

Over and above all, we shall try to give the boy some stimulus for his work. Unless he takes an interest in what he is doing, he will p. 5 never become an artisan in the true sense of the word.

Go through the book, and see whether, here and there, you do not get some glimpses of what it means to take a pleasure in doing each particular thing, and you will find in every instance that it is a satisfaction because you have learned to perform it with ease.

I do not know of anything which has done as much to advance the arts and manufactures, during the last century, as the universal desire to improve the form, shape and structure of tools; and the effort to invent new ones. This finds its reflection everywhere in the production of new and improved products.

In this particular I have been led to formulate a homely sentence which expresses the idea: Invention consists in doing an old thing a new way; or a new thing any way.

The Author.

p. 7

CHAPTER I ToC

ON TOOLS GENERALLY

Judging from the favorable comments of educators, on the general arrangement of the subject matter in the work on "Carpentry for Boys," I am disposed to follow that plan in this book in so far as it pertains to tools.

In this field, as in "Carpentry," I do not find any guide which is adapted to teach the boy the fundamentals of mechanics. Writers usually overlook the fact, that as the boy knows nothing whatever about the subject, he could not be expected to know anything about tools.

To describe them gives a start in the education, but it is far short of what is necessary for one in his condition. If he is told that the chisel or bit for a lathe has a diamond point, or is round-nosed, and must be ground at a certain angle, he naturally wants to know, as all boys do, *why* it should be at that angle.

So in the setting of the tools with relation to the work, the holding and manipulation of the file, of the drill for accurate boring, together with numerous little things, are all taken for granted, and the boy blunders along with the ultimate ob p. 8 ject in sight, without having the pathway cleared so he may readily reach the goal.

Varied Requirements.—The machinist's trade is one which requires the most varied tools of all occupations, and they are by all odds the most expensive to be found in the entire list of vocations.

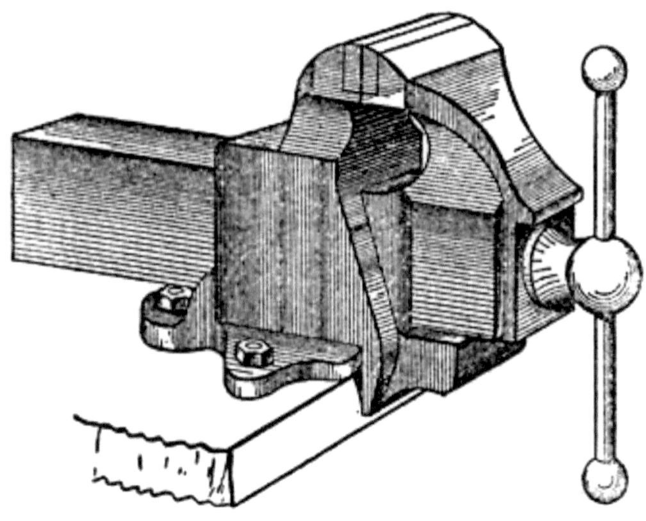

1. Bench Vise. ToList

This arises from the fact that he must work with the most stubborn of all materials. He finds resistance at every step in bringing forth a product.

List of Tools. — With a view of familiarizing p. 9 the boy with this great variety the following list is compiled, from which we shall select the ones essential in the initial equipment of a small shop.

Vises. — One small, good vise is infinitely preferable to two bad ones. For ordinary work a 3-inch jaw is preferable, and it should be firmly mounted on the bench. So many kinds are now made that it would be a costly thing to purchase one for each special use, therefore the boy will find it profitable to make some attachments for the ordinary vise.

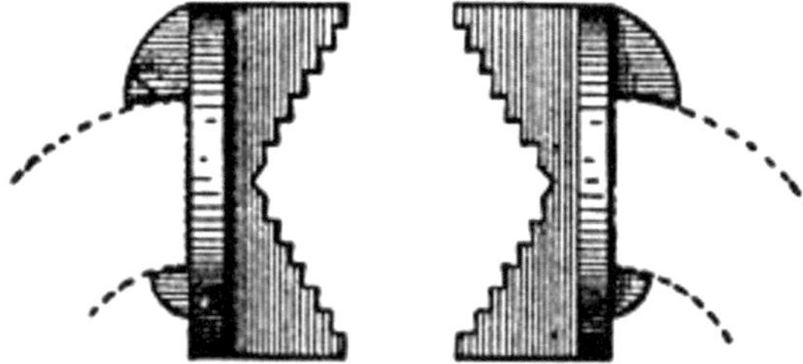

Fig. 2.

Pipe Grip for Vise. ToList

Swivel Vises.—A swivel vise is always a good tool, the cost being not excessive over the ordinary kind. Then a pair of grips for holding pipe, or round material which is to be threaded, can readily be made.

The drawing (Fig. 2) shows a serviceable pair of grips, made to fit the jaws of a vise, and will p. 10 be acceptable in much of the work. Then, the vise should be provided with copper caps for the jaws to be used when making up articles which would otherwise be injured by the jaws.

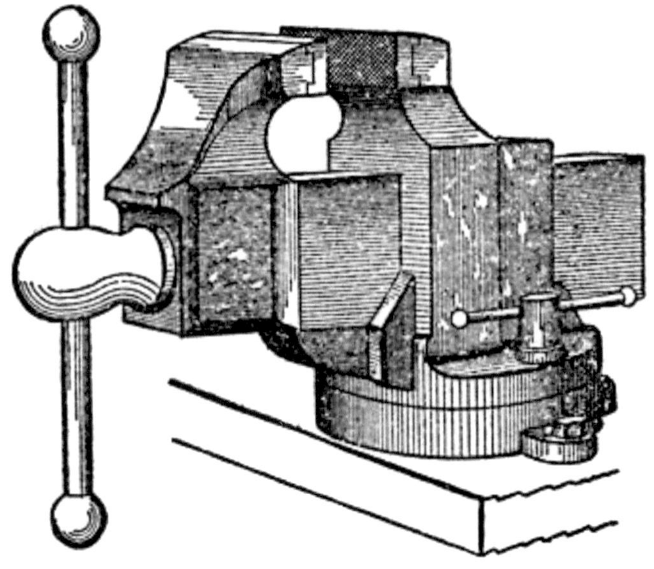

3. Swivel Vise. ToList

Let us get a comprehensive view of the different kinds of tools necessary in a fully equipped shop.

Parts of Lathe.—The first thing of importance is the lathe, and of these there is quite a variety, and among the accompaniments are the slide rest, mandrel, back gear, division plate, angle plate, cone plate and various chucks

p. 11

There must also be change wheels, studs and quadrant plates, self-acting feed for surfacing and cross slide, and clamping nuts.

Drilling machines, both hand and power, hand and ratchet braces and breast-drill stocks.

Fig.

4. — Speed Lathe. ToList

Chisels. — Chisels of various kinds, for chipping and cross-cutting; round-nosed, centering, set punches, tommies and drifts.

Back, tee and centering square; bevels, spirit level, inside and out-side calipers, straight edges, rules and surface plates

p. 12

Gages for boring, scribing blocks, steel and brass scribes, stocks and dies, screw-plates, taps for bolts, reamers.

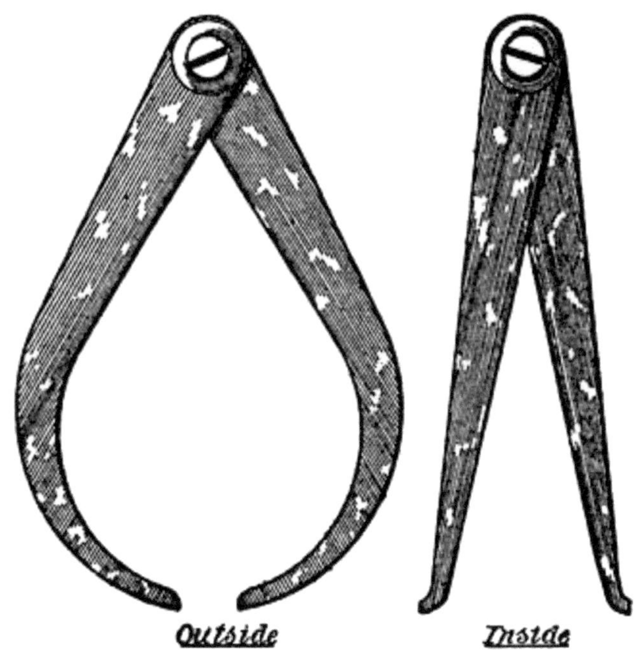

Fig. 5.

Calipers ToList

Files for various descriptions, countersinks, frame and hack saws.

Grinding Apparatus.—Emery wheel, cloth and paper, paper, flour emery, polishing powders, laps and buffs, and polishing sticks

p. 13

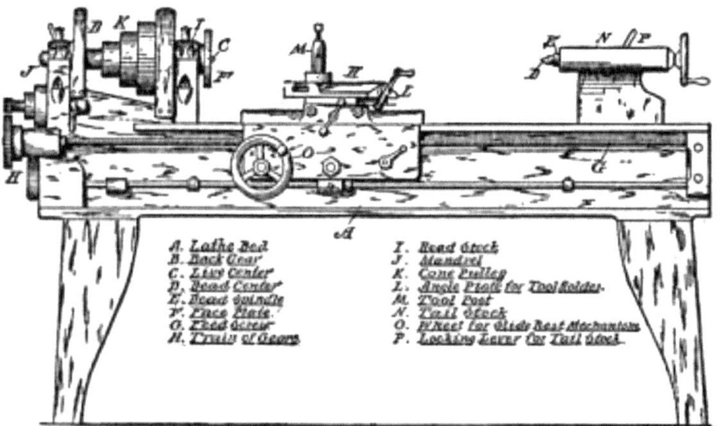

Fig. 6. Engine Lathe. ToList
p. 14

Forge, anvils, tongs, swages, punches, bolt tools, hot and cold chisels, blow-pipe, soldering iron, hard and soft solders, borax, spirits of salts, oil, resin and spelter.

To this may be added an endless variety of small bench tools, micrometers, protractors, arbors, collets, box tools and scrapers.

Fig. 7.

Center Gage. ToList

Large Machines. — The list would not be complete without the planer, shaper and milling machine, with their variety of chucks, clamps and other attachments, too numerous to mention.

The foregoing show what a wonderful variety of articles are found in a well-equipped shop, all of which can be conveniently used; but to the boy who has only a small amount of money, a workable set is indicated as follows:

A small lathe, with an 8-inch swing, can be obtained at a low cost, provided with a countershaft complete.

Chucks.—With this should go a small chuck, p. 15 and a face-plate for large work, unless a large chuck can also be acquired. This, with a dozen tools of various sizes, and also small bits for drilling purposes.

The lathe will answer all purposes for drilling, but small drilling machines are now furnished at very low figures, and such a machine will take off a great deal of duty from the lathe.

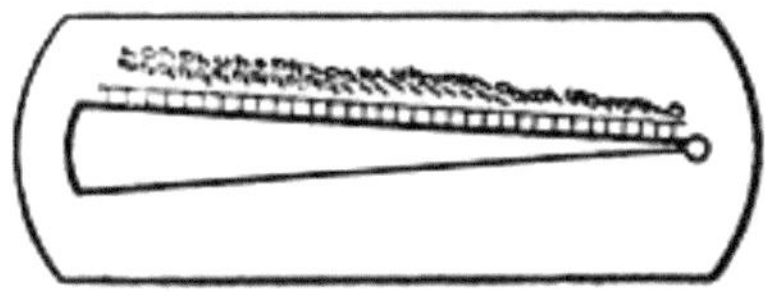

Fig. 8.

Pocket Screw and Wire Gage. ToList

As the lathe is of prime importance, never use it for drilling, if you have a driller, as it always has enough work to do for tuning up work.

Bench Tools.—Of bench tools, a 3-inch vise, various files, center punch, two hammers, round and A-shaped peons, hack saw, compasses, inside and outside calipers, screw driver, cold chisels, metal square, level, straight edge, bevel square, reamers, small emery wheel and an oil stone, make a fairly good outfit to start with, and these can be added to from time to time.

Everything in the machine shop centers about p. 16 the lathe. It is the king of all tools. The shaper and planer may be most efficient for surfacing, and the milling machine for making grooves and gears, or for general cutting purposes, but the lathe possesses a range of work not possible with either of the other tools, and for that reason should be selected with great care.

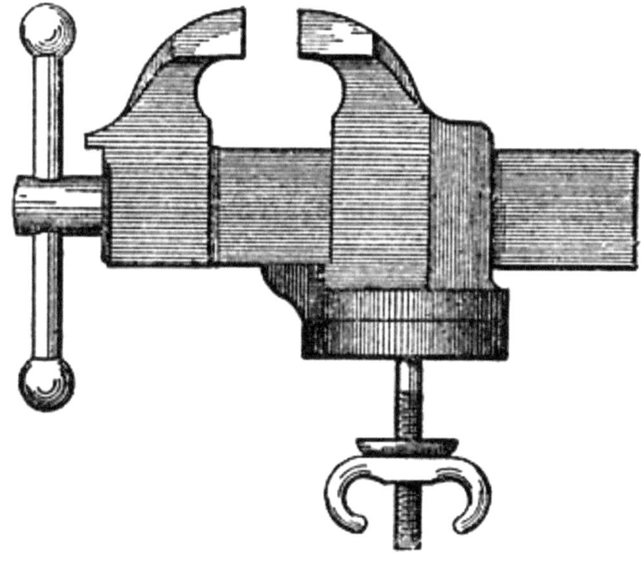

Fig. 9.

Handy Bench Vise. ToList

Selecting a Lathe.—The important things about a lathe are the spindle bearings and the ways for the tool-holder. The least play in either will ruin any work. Every other part may be p. 17 defective, but with solidly built bearing-posts and bearings, your lathe will be effective.

For this reason it will not pay to get a cheap tool. Better get a small, 6-inch approved tool of this kind, than a larger cheap article. It may pay with other tools, but with a lathe never.

Never do grinding on a lathe. The fine emery, or grinding material, is sure to reach the bearings; it matters not what care is exercised. There is only one remedy for this—overhauling.

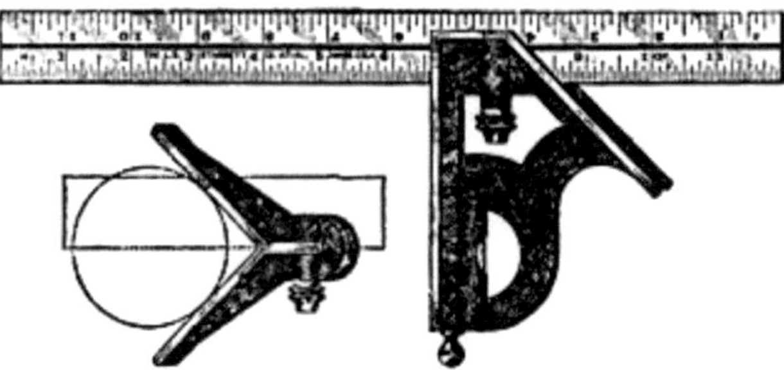

10. — Combination Square. ToList

Combination Square. — A tool of this kind is most essential, however small. It can be used as a try-square, and has this advantage, that the head can be made to slide along the rule and be clamped at any point. It has a beveling and a leveling device, as well.

Fig. 11.—Uses of the Combination Square. ToList

The combination square provides a means for p. 19 doing a great variety of work, as it combines the qualities of a rule, square, miter, depth gage, height gage, level and center head.

12. — A Quick Adjusting Micrometer. ToList

The full page illustration (Fig. 11) shows some of the uses and the particular manner of holding the tool.

Micrometers. — Tools of this description are made which will accurately measure work in di p. 20 mensions of ten-thousandths of an inch up to an inch.

The illustration (Fig. 12) shows an approved tool, and this is so constructed that it can instantly be changed and set by merely pressing the end of the plunger as shown.

Fig. 13. — A Universal Bevel Protractor. ToList

Protractors. — As all angles are not obtainable by the square or bevel, a protractor is a most desirable addition to the stock of tools. As one side of the tool is flat it is convenient for laying on the paper when drafting, as well as for use on the work.

The protractor has a graduated disk, and is adjustable so it can be disposed at any angle.

p. 21

Fig.
14.—Universal Bevel Protractor,
showing its uses. ToList
p. 22

All special tools of this kind are serviceable, and the boy should understand their uses, even though he is not able for the time being to acquire them. To learn how they are applied in daily use is an education in itself.

Utilizing Bevel Protractor.—Examine the full-page illustration (Fig. 14), and see how the bevel protractor is utilized to measure the angles of work, whether it is tapering heads or different kinds of nuts, or end and side surfacing, and it will teach an important lesson.

Fig. 15.—Grindstone Truing Device. ToList

Truing Grindstones.—Devices for truing up grindstones are now made, and the illustration (Fig. 15) shows a very efficient machine for this purpose. It can be applied instantly to the face p. 23 of the stone, and it works automatically, without interfering with the use of the stone.

It is frequently the case that an emery wheel will become glazed, due to its extreme hardness. This is also caused, sometimes, by running it at too high a speed. If the glazing continues after the speed is reduced, it should be ground down an eighth of an inch or so. This will, usually, remedy the defect.

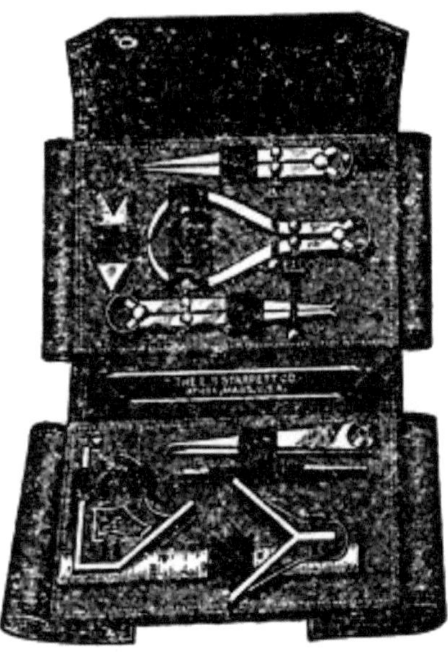

Fig. 16.—Set of Tools and Case. ToList

Sets of Tools.—A cheap and convenient set of p. 24 precision tools is shown in Fig. 16, which is kept in a neat folding leather case. The set consists of a 6-inch combination square, complete center punch, 6-inch flexible steel rule center gage, 4-inch calipers, 4-inch outside caliper with solid nut, 4-inch inside caliper with solid nut, and a 4-inch divider with a solid nut.

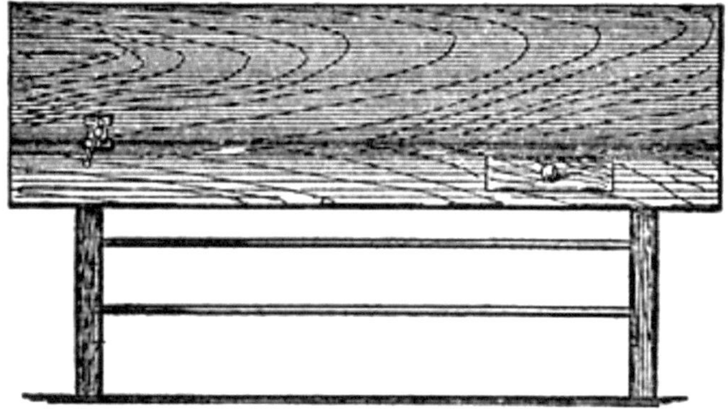

Fig.

17. The Work Bench. ToList

The Work Bench.—This is the mechanic's fort. His capacity for work will depend on its arrangement. To the boy this is particularly interesting, and for his uses it should be made full three inches lower than the standard height.

A good plan to judge of the proper height is to measure from the jaws of the vise. The top of the jaw should be on a level with the elbows. Grasp a file with both hands, and hold it as though p. 25 in the act of filing across the work; then measure up from the floor to the elbows, when they are held in that position.

The Proper Dimensions.—This plan will give you a sure means of selecting a height that is best adapted for your work. The regulation bench is about 38 inches high, and assuming that the vise projects up about 4 inches more, would bring the top of the jaws about 42 to 44 inches from the floor. It is safe to fix the height of the bench at not less than 34 inches.

This should have a drawer, preferably near the right-hand end of the bench. The vise should be at the left side, and the bench in your front should be free of any fixed tools.

How Arranged.—Have a rack above the bench at the rear, for the various tools when not in use, and the rear board of the bench should be elevated above the front planks several inches, on which the various tools can be put, other than those which are suspended on the rack above.

The advantage of this is, that a bench will accumulate a quantity of material that the tools can hide in, and there is nothing more annoying than to hunt over a lot of trash to get what is needed. It is necessary to emphasize the necessity of always putting a tool back in its proper place, immediately after using

p. 26

CHAPTER II ToC

HOW TO GRIND AND SHARPEN TOOLS

It is singular, that with the immense variety of tools set forth in the preceding chapter, how few, really, require the art of the workman to grind and sharpen. If we take the lathe, the drilling machine, as well as the shaper, planer, milling machine, and all power-driven tools, they are merely mechanism contrived to handle some small, and, apparently, inconsequential tool, which does the work on the material.

Importance of the Cutting Tool.—But it is this very fact that makes the preparation of that part of the mechanism so important. Here we have a lathe, weighing a thousand pounds, worth hundreds of dollars, concentrating its entire energies on a little bit, weighing eight ounces, and worth less than a dollar. It may thus readily be seen that it is the little bar of metal from which the small tool is made that needs our care and attention.

This is particularly true of the expensive milling machines, where the little saw, if not in perfect order, and not properly set, will not only do improper work, but injure the machine itself. p. 27 More lathes are ruined from using badly ground tools than from any other cause.

In the whole line of tools which the machinist must take care of daily, there is nothing as important as the lathe cutting-tool, and the knowledge which goes with it to use the proper one.

Let us simplify the inquiry by considering them under the following headings:

1. The grinder.

2. The grinding angle.

The Grinder.—The first mistake the novice will make, is to use the tool on the grinder as though it were necessary to grind it down with a few turns of the wheel. Haste is not conducive to proper sharpening. As the wheel is of emery, corundum or other quickly cutting material, and is always run at a high rate of speed, a great heat is evolved, which is materially increased by pressure.

Pressure is injurious not so much to the wheel as to the tool itself. The moment a tool becomes heated there is danger of destroying the temper, and the edge, being the thinnest, is the most violently affected. Hence it is desirable always to have a receptacle with water handy, into which the tool can be plunged, during the process of grinding down.

Correct Use of Grinder. — Treat the wheel as p. 28 though it is a friend, and not an enemy. Take advantage of its entire surface. Whenever you go into a machine shop, look at the emery wheel. If you find it worn in creases, and distorted in its circular outline, you can make up your mind that there is some one there who has poor tools, because it is simply out of the question to grind a tool correctly with such a wheel.

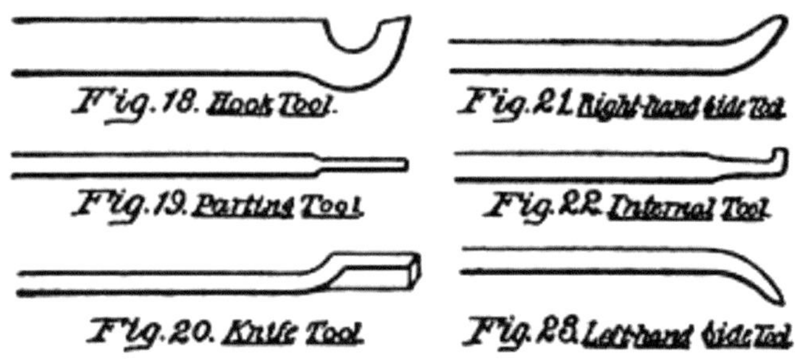

Fig. 18-23. Tools. ToList

Coarse wheels are an abomination for tool work. Use the finest kinds devised for the purpose. They will keep in condition longer, are not so liable to wear unevenly, and will always finish off the edge better than the coarse variety.

Lathe Bits. — All bits made for lathes are modifications of the foregoing types (Figs. 18-23, 19, 20, 21, 22, 23).

As this chapter deals with the sharpening methods only, the reader is referred to the next chap p. 29 ter, which deals with the manner of setting and holding them to do the most effective work.

When it is understood that a cutting tool in a lathe is simply a form of wedge which peels off a definite thickness of metal, the importance of proper grinding and correct position in the lathe can be appreciated.

Roughing Tools.—The most useful is the roughing tool to take off the first cut. As this type of tool is also important, with some modifications, in finishing work, it is given the place of first consideration here.

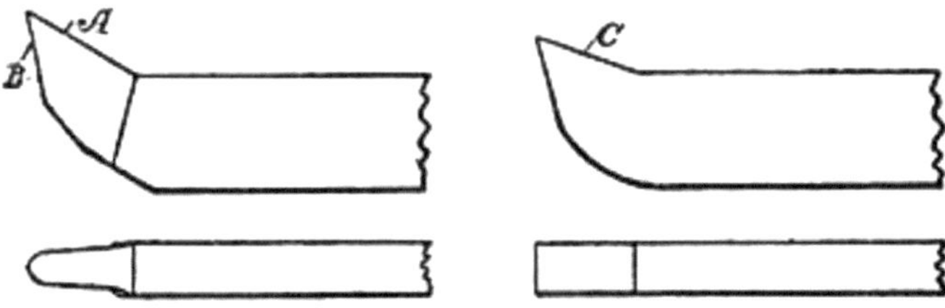

Fig. 24. Tool for Wrought Iron.   Fig. 25. Tool for Cast Iron. ToList

Fig. 24 shows side and top views of a tool designed to rough off wrought iron, or a tough quality of steel. You will notice, that what is called the top rake (A) is very pronounced, and, as the point projects considerably above the body of the tool itself, it should, in practice, be set with its cutting point above the center.

p. 30

The Clearance.—Now, in grinding, the important point is the clearance line (B). As shown in this figure, it has an angle of 10 degrees, so that in placing the tool in the holder it is obvious it cannot be placed very high above the center, particularly when used on small work. The top rake is ground at an angle of 60 degrees from the vertical. The arc of the curved end depends on the kind of lathe and the size of the work.

The tool (Fig. 25), with a straight cutting edge, is the proper one to rough off cast iron. Note that the top rake (C) is 70 degrees, and the clearance 15 degrees.

The Cutting Angle.—Wrought iron, or mild steel, will form a ribbon when the tool wedges its way into the material. Cast iron, on the other hand, owing to its brittleness, will break off into small particles, hence the wedge surface can be put at a more obtuse angle to the work.

In grinding side-cutters the clearance should be at a less angle than 10 degrees, rather than more, and the top rake should also be less; otherwise the tendency will be to draw the tool into the work and swing the tool post around.

Drills.—Holders for grinding twist drills are now furnished at very low prices, and instructions are usually sent with the machines, but a few words may not be amiss for the benefit of those p. 31 who have not the means to purchase such a machine.

Hand grinding is a difficult thing, for the reason that through carelessness, or inability, both sides of the drill are not ground at the same angle and pitch. As a result the cutting edge of one side will do more work than the other. If the heel angles differ, one side will draw into the work, and the other resist.

Fig. 26. End view.   Fig. 27. Side view. ToList

Wrong Grinding.—When such is the case the hole becomes untrue. The sides of the bit will grind into the walls, or the bit will have a tendency to run to one side, and particularly if boring through metal which is uneven in its texture or grain.

Figs. 26 and 27 show end and side views of a bit properly ground. If a bit has been broken off, first grind it off square at the end, and then grind down the angles, so that A is about 15 degrees, and be sure that the heel has sufficient p. 32 clearance—that is, ground down deeper than the cutting point.

Chisels. — A machine shop should always have a plentiful supply of cold chisels, and a particular kind for each work, to be used for that purpose only. This may seem trivial to the boy, but it is really a most important matter.

Notice the careless and incompetent workman. If chipping or cutting is required, he will grasp the first chisel at hand. It may have a curved end, or be a key-way chisel, or entirely unsuited as to size for the cutting required.

The result is an injured tool, and unsatisfactory results. The rule holds good in this respect as with every other tool in the kit. *Use a tool for the purpose it was made for*, and for no other. Acquire that habit.

Cold Chisels. — A cold chisel should never be ground to a long, tapering point, like a wood chisel. The proper taper for a wood chisel is 15 degrees, whereas a cold chisel should be 45 degrees. A drifting chisel may have a longer taper than one used for chipping.

It is a good habit, particularly as there are so few tools which require grinding, to commence the day's work by grinding the chisels, and arranging them for business.

System in Work. — Then see to it that the drills p. 33 are in good shape; and while you are about it, look over the lathe tools. You will find that it is better to do this work at one time, than to go to the emery wheel a dozen times a day while you are engaged on the job.

Adopt a system in your work. Don't take things just as they come along, but form your plans in an orderly way, and you will always know how to take up and finish the work in the most profitable and satisfactory way.

Wrong Use of Tools. — Never use the vise as an anvil. Ordinary and proper use of this tool will insure it for a lifetime, aside from its natural wear. It may be said with safety that a vise will never break if used for the purpose for which it was intended. One blow of a hammer may ruin it.

Furthermore, never use an auxiliary lever to screw up the jaws. If the lever which comes with it is not large enough to set the jaws, you may be sure that the vise is not large enough for your work

p. 34

CHAPTER III ToC

SETTING AND HOLDING TOOLS

Some simple directions in the holding and setting of tools may be of service to the novice. Practice has shown the most effective way of treating different materials, so that the tools will do the most efficient work.

A tool ground in a certain way and set at a particular angle might do the work admirably on a piece of steel, but would not possibly work on aluminum or brass.

Lathe Speed.—If the lathe should run at the same speed on a piece of cast iron as with a brass casting, the result would not be very satisfactory, either with the tool or on the work itself.

Some compositions of metal require a high speed, and some a hooked tool. These are things which each must determine as the articles come to the shop; but there are certain well-defined rules with respect to the ordinary metals that should be observed.

The Hack Saw.—Our first observation should be directed to the hand tools. The hack saw is one of the most difficult tools for the machinist to handle, for the following reasons: p. 35

First, of the desire to force the blade through the work. The blade is a frail instrument, and when too great a pressure is exerted it bends, and as a result a breakage follows. To enable it to do the work properly, it must be made of the hardest steel. It is, in consequence, easily fractured.

Fig. 28.—Hack Saw Frame.
Fig. 29.—Hack Saw Blade. ToList

Second. The novice will make short hacking cuts. This causes the teeth to stick, the saw bends, and a new blade is required. Take a long sweeping cut, using the entire length of the blade. Do not oscillate the blade as you push it through the work, but keep the tooth line horizontal from one end of the stroke to the other. The moment it begins to waver, the teeth will catch on the metal on the side nearest to you, and it will snap

p. 36

Third. The handle is held too loosely. The handle must be firmly held with the right hand, and the other held by the fingers lightly, but in such a position that a steady downward pressure can be maintained. If loosely held, the saw is bound to sag from side to side during the stroke, and a short stroke accentuates the lateral movement. A long stroke avoids this.

The hack saw is one of the tools which should be used with the utmost deliberation, combined with a rigid grasp of the handle.

Files.—For remarks on this tool see Chapter IV, which treats of the subject specially.

Grindstones, Emery and Grinding Wheels.—A good workman is always reflected by his grinding apparatus. This is true whether it has reference to a grindstone, emery, corundum wheel, or a plain oil stone. Nothing is more destructive of good tools than a grooved, uneven, or wabbly stone. It is only little less than a crime for a workman to hold a tool on a revolving stone at one spot.

Carelessness in Holding Tools.—The boy must learn that such a habit actually prevents the proper grinding, not only of the tool he has on the stone, but also of the one which follows. While it is true that all artificially made grinders will wear unevenly, even when used with the utmost care, due to uneven texture of the materials in the stone, still, p. 37 the careless use of the tool, while in the act of grinding, only aggravates the trouble.

Another fault of the careless workman is, to press the bit against the stone too hard. This cuts the stone more than it wears off the tool, and it is entirely unnecessary. Furthermore, it heats up the tool, which should be avoided.

Calipers.—A true workman, who endeavors to turn out accurate work, and preserve his tools, will never test the work with his calipers while the piece is turning in the lathe. A revolving cast iron disk will cut ruby, the hardest substance next to the diamond, so it is not the hardness of the material which resists wear, but the conditions under which it is used.

Care in Use of Calipers.—The calipers may be of the most hardened steel, and the work turned up of the softest brass, the latter, when revolving, will grind off the point of the tool, for the reason that the revolving piece constantly presents a new surface to the point of the calipers, and when tests are frequently made, it does not take long to change the caliper span so that it must be reset.

As stated elsewhere, the whole energy of the lathe is concentrated on the bit or cutting tool, hence, in order to get the most effective work out of it requires care; first, in grinding; and, second, in setting

p. 38

Machine Bits.—It does not always matter so much whether you use a square, pointed, or a round-nosed bit, provided it is properly ground and set in the tool holder. As a rule, the more brittle the metal the less the top rake or angle should be.

In the chapter relating to the grinding of tools, references were made as to the most serviceable bits for the various metals. We are concerned here with the setting or holding of these articles.

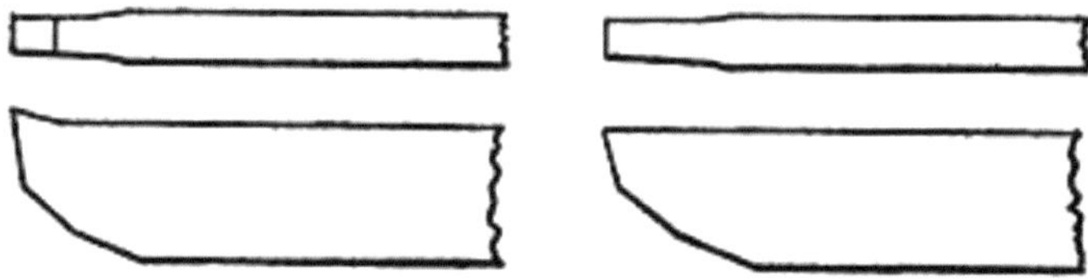

Fig. 30. Plain Hook Tool   Fig. 31. Plain Straight Tool ToList

The two illustrations here given show a pair of plain bits, in which Fig. 30 represents a hook-shaped formation, and Fig. 31 a straight grind, without any top rake. The hooked bit would do for aluminum, or steel, but for cast iron the form shown in Fig. 31 would be most serviceable.

Then the side bits, such as the round-nosed, Fig. 32 and the square end, Fig. 33, may be ground hooked, or with a top rake, or left flat.

The too common mistake is to grind the lower or clearance side at too great an angle. Fig. 34 p. 39 shows the correct angle, and the dotted line A illustrates the common tendency to grind the clearance.

The Proper Angle for Lathe Tools.—Now there is a reason why the angle of from 10 to 15 should be maintained in the clearance. The point of the tool must have suitable support for the work it is required to do, so it will not chatter or yield in the slightest degree. A bit ground along the dotted line has a cutting edge which will spring down, and consequently break or produce a rough surface.

50

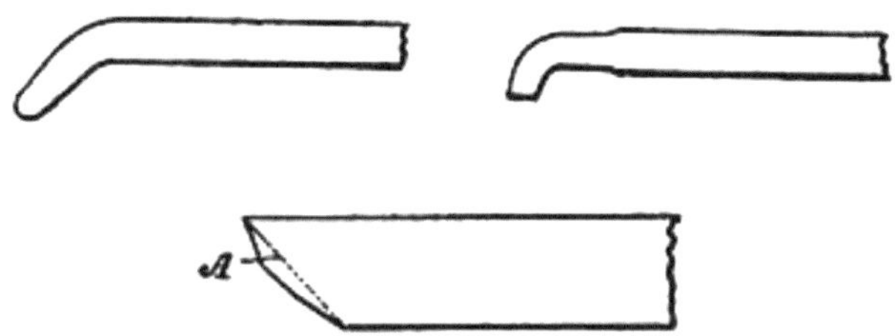

Figs.
32, 33, 34.
Proper Angles for Tools ToList

Then, again, the angle of the clearance acts as a guide, or rather, a guard, to prevent the tool from going in too far, as will now be explained.

Setting the Bit.—In order to understand the correct setting, examine the work A, in Fig. 35.

A is a cylinder being turned up in the lathe, and B the cutting tool, which approaches it on a hori p. 40 zontal line, C, extending out from the center of the cylinder A. This setting is theoretically correct, and in practice has been found most advantageous.

In this case let us assume that the clearance angle D is 15 degrees, as well as in the following figures.

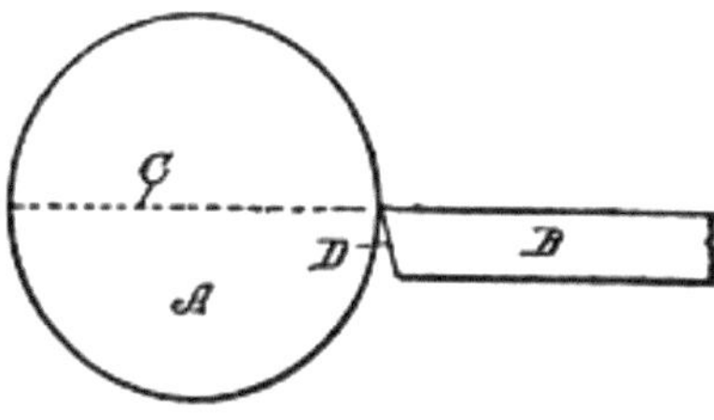

Fig. 35.

Set of the Bit ToList

Suppose we have a piece of tough steel, and the tool holder is raised so that the point of the tool is at the 15 degree line E, as shown in Fig. 36, in which case the clearance line D is at right angles to the line E. The line E is 15 degrees above the center line C.

The Setting Angle.—Now, it is obvious that if the tool should be raised higher than the line E it would run out of work, because the clearance surface of the tool would ride up over the surface cut by the edge of the tool.

If, on the other hand, the tool should be placed lower, toward the line C, the tendency would be to draw in the tool toward the center of the work A

p. 41

In Fig. 37 the tool has its point elevated, in which case it must be lowered so the point will touch the work nearer the center line C.

The foregoing arrangement of the tools will be found to be effective where the material is soft and not too tough as with aluminum.

Bad Practice.—Figs. 38 and 39 show illustrations of bad practice which should never be resorted to. Fig. 38 shows the tool, held in a horizontal position, but with its point below the center line C. With any rough metal the tool could not possibly work, except to act as a scraper, and if it should be used in that position on cast iron, the tool itself would soon be useless.

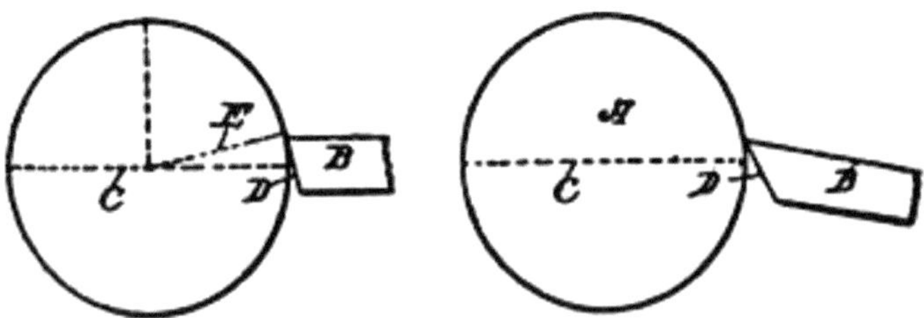

Fig. 36. Correct Angle   Fig. 37. Wrong Angle ToList

Fig. 39 is still worse, and is of no value for any purpose except in polishing brass, where it would be serviceable. It would make a sorry looking job with aluminum. Brass requires a tool with very p. 42 little top rake, and the point should be set near the center line C.

Lathe Speed.—It is often a question at what speeds to run the lathe for different work. If you know the speeds of your lathe at low

52

and high gear, you must also consider the diameter of the work at the cutting point.

The rule is to have the bit cut from 15 to 20 feet per minute for wrought iron; from 11 to 18 feet for steel; from 25 to 50 for brass; and from 40 to 50 for aluminum.

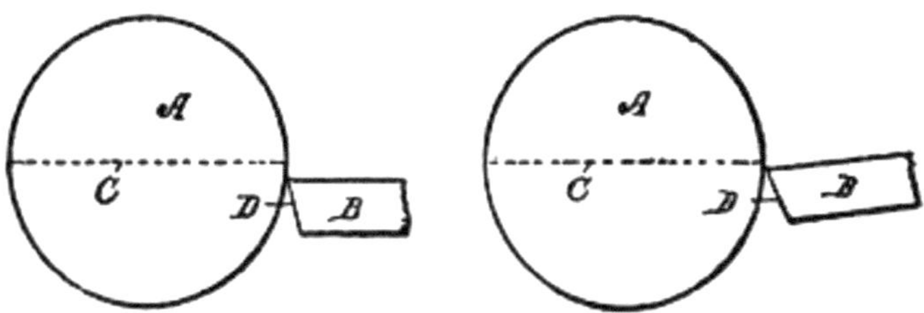

Fig. 38. Too Low   Fig. 39. Improper Set ToList

As a result, therefore, if, at low speed, a piece 10 inches in diameter, runs at the proper speed to cut at that distance from the center, it is obvious that a piece 5 inches in diameter should ran twice as fast. This is a matter which time and practice will enable you to judge with a fair degree of accuracy

p. 43

Observe this as a maxim: "Slow speed, and quick feed."

Boring Tools on Lathe. — The lathe is a most useful tool for boring purposes, better for some work than the drilling machine itself. The work which can be done better on a lathe than on a drilling machine, may be classified as follows:

1. When straight and true holes are required.

2. In long work, where the lathe is used to turn up the article, and where the drilling can be done at the same time.

3. Anything that can be chucked in a lathe.

4. Where the work is long and cannot be fixed in a drilling machine. The long bed of the lathe gives room for holding such work.

Fig. 40. Internal Set   Fig. 41. Set for Brass ToList

The Rake of the Drill.— A boring tool requires some knowledge in setting. It should have a greater top rake than for the outside work, and the cutting edge should also be keener, as a rule

p. 44

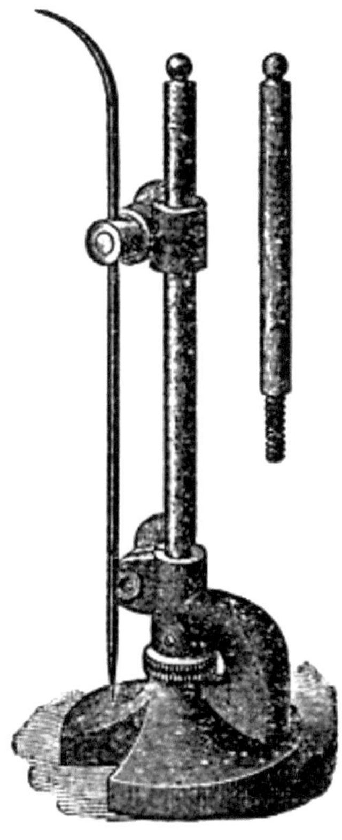

42. — Surface Gage. ToList

In this class of work the material bored must be understood, as well as in doing outside work.

The hooked tool, Fig. 40, is shown to be considerably above the center line, and at that point it will do the most effective cutting on steel. If, on the other hand, brass is operated on there should be no p. 45 top rake, as illustrated in Fig. 41, thus assuring a smooth job.

Laps. — This is a tool which is very useful, particularly for grinding and truing up the cylinders of internal combustion engines, as well as for all kinds of bores of refractory material which cannot be handled with the cutting tool of the lathe.

It is made up of a mandrel or rod of copper, with lead cast about it, and then turned up true, so that it is but the merest trifle larger than the hole it is to true up.

Using the Lap. — The roller thus made is turned rapidly in a lathe, and the cylinder to be trued is brought up to it and the roller supplied freely with emery powder and oil. As rapidly as possible the cylinder is worked over on the roller, without forcing it, and also turned, so as to prevent even the weight from grinding it unduly on one side.

More or less of the emery will embed itself in the lead, and thus act as an abrasive. The process is called "lapping."

Surface Gages. — Frequently, in laying out, it is necessary to scribe lines at a given distance from some part of the work; or, the conditions are such that a rule, a caliper, or dividers will not permit accurate measurement to be made.

For such purposes, what is called a surface gage was devised. This is merely a heavy base, provided p. 46 with a pivoted upright on which is mounted a scribe that is held by a clamp so it may be turned to any angle.

p. 47

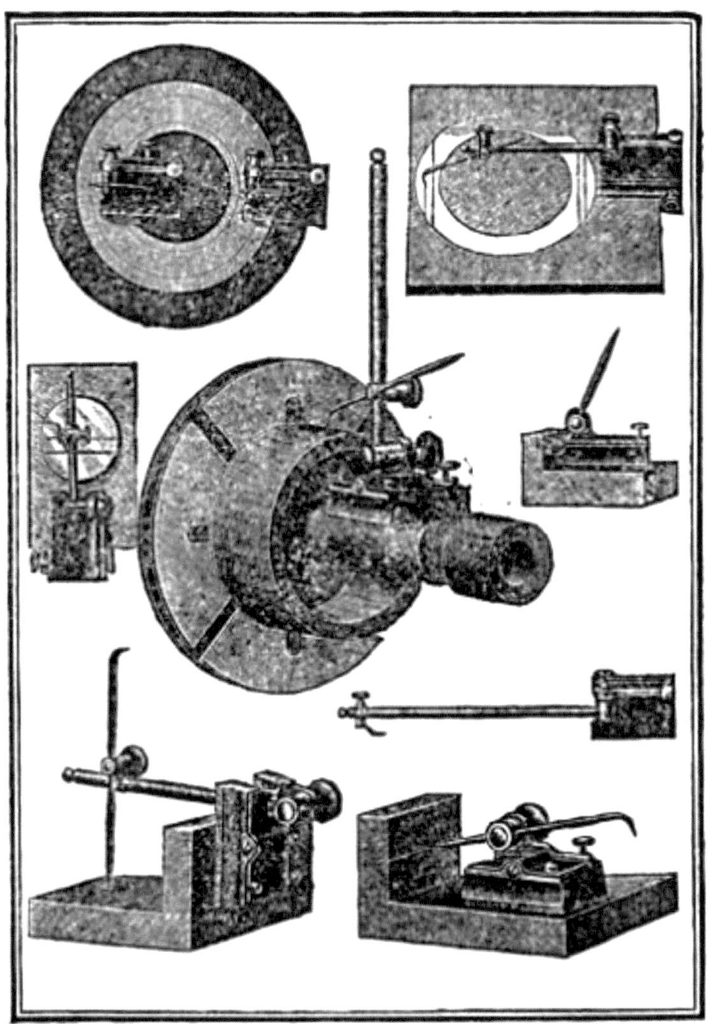

Fig. 43.—Showing uses of the Surface Gage. ToList

Surface Gage.—The clamp holding the scriber is vertically movable on the pivoted upright. By resting the base of the surface gage
on the line to be measured from, and swinging one point of the
scriber to the place where the work is to be done, accuracy is as-

sured. One end of the scriber is bent, so it can be adapted to enter recesses, or such places as could not be reached by the straight end

p. 48

CHAPTER IV ToC

ON THE USE OF THE FILE

The most necessary tool in a machine shop is a file. It is one of the neglected tools, because the ordinary boy, or workman, sees nothing in it but a strip or a bar with a lot of cross grooves and edges, and he concludes that the only thing necessary is to rub it across a piece of metal until he has worn it down sufficiently for the purpose.

The First Test.—The fact is, the file is so familiar a tool, that it breeds contempt, like many other things closely associated in life.

Give the boy an irregular block of metal, and tell him to file it up square, and he will begin to realize that there is something in the handling of a file that never before occurred to him.

He will find three things to astonish him:

First: That of dimensions.

Second: The difficulty of getting it square.

Third: The character of the surface when he has finished it.

Filing an Irregular Block.—To file a block of an irregular character so that the dimensions are accurate, is a good test for an accomplished workman. The job is made doubly difficult if he is required to file it square at the same time. It will p. 49 be found, invariably, that the sides will not be parallel, and by the time it is fully trued up the piece will be too small. See Figs. 44 and 45.

Then, unless the utmost care is taken, the flat sides *will not* be flat, but rounded.

Filing a Bar Straight.—The next test is to get the boy to file a bar straight. He has no shaper or planer for the purpose, so that it must be done by hand. He will find himself lacking in two things: The edge of the bar will not be straight; nor will it be square with the side of the bar.

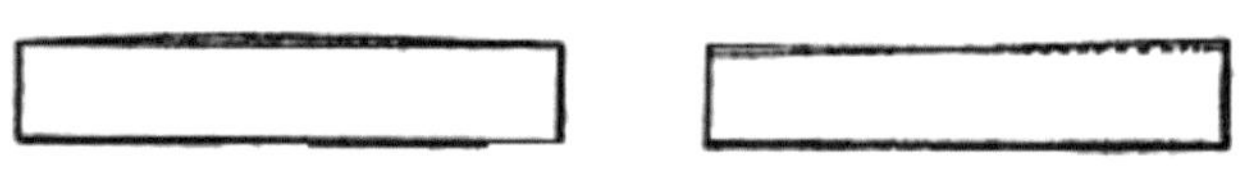

Fig. 44. Rounded Surface   Fig. 45. A Winding Face ToList

Filing Bar with Parallel Sides.—Follow up this test by requiring him to file up a bar, first, with two exactly parallel sides, and absolutely straight, so it will pass smoothly between the legs of a pair of calipers, and then file the two other sides in like manner.

Surfacing off Disks.—When the foregoing are completed there is still another requirement which, though it appears simple, is the supreme test. Set him to work at surfacing off a pair of disks or plates, say one and a half inches in diameter, so p. 50 that when they are finished they will fit against each other perfectly flat.

A pair of such disks, if absolutely true, will hold together by the force of cohesion, even in a dry state, or they will, as it were, float against each other.

True Surfacing.—Prior to about 1850 the necessity of true surfacing was not so important or as well known as at the present time. About that period Sir J. Whitworth, an eminent English engineer and mechanic, called the attention of machinists to the great advantage arising from true surfaces and edges for all types of machinery, and he laid the foundation of the knowledge in accurating surfacing.

Precision Tools.—Due to his energy many precision tools were made, all tending to this end, and as a result machines became better and more efficient in every way.

It had this great advantage: It taught the workman of his day how to use the file and scraper, because both must be used conjunctively to make an absolutely flat plate.

Contrary to general beliefs, shapers and planers do not make absolutely accurate surfaces. The test of this is to put together two plates so planed off. There is just enough unevenness to permit air to get between the plates. If they were perfectly true p. 51 they would exclude all air, and it would be a difficult matter to draw them apart.

Test of the Mechanic. — To make them perfectly flat, one plate has chalk rubbed over it, and the two plates are then rubbed together. This will quickly show where the high spots are, and the file and scraper are then used to cut away the metal.

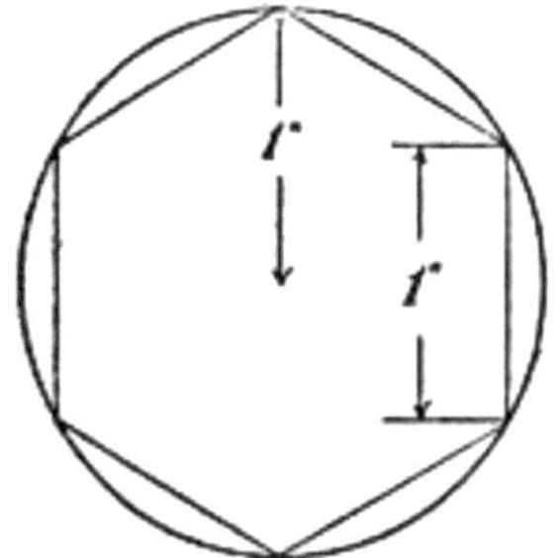

Fig. 46. Hexagon Nut

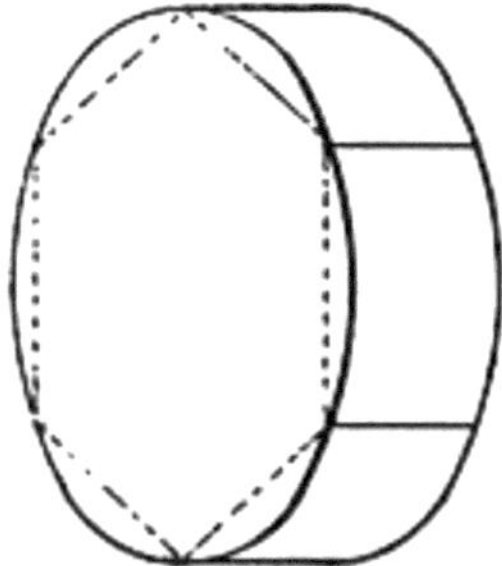

Fig. 47. Hexagon Nut ToList

In England the test of the mechanic used to be determined by his ability to file a piece of metal flat. It was regarded as the highest art. This is not the most desirable test at the present time, and it is recognized that a much severer test is to file a narrow piece exactly flat, and so that it will not have a trace of roundness, and be square from end to end.

Test Suggestions. — In a shop which does not have the advantage of a planer or shaper, there p. 52 are so many articles which must be filed up, that it is interesting to know something of how the various articles are made with a file.

To file a hexagon, or six-sided nut will be a good test with a file. To do this a little study in geometrical lines will save a vast amount of time. In beginning the work, measure the radius with a divider, and then step off and make six marks equidistant from each other on the round surface.

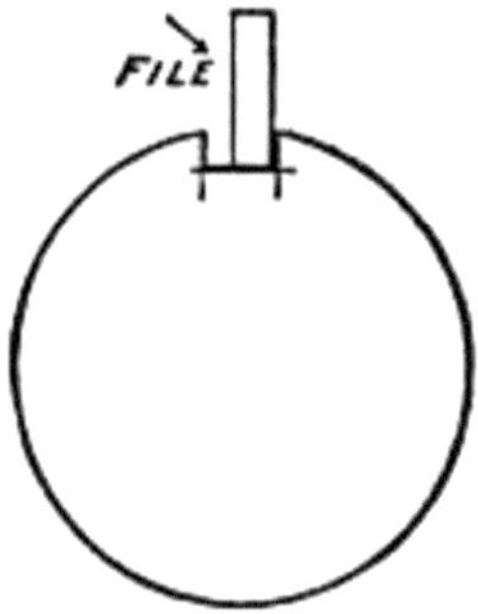

Fig.

48. Cutting Key-way ToList

Use of the Dividers.—The distance between each of these points is equal to the radius, or half the diameter, of the round bar. See Fig. 46, which shows this. The marks should be scribed across the surface, as shown in Fig. 47, where the lines show the ends of the facets of the outside of the nut.

Do not let the file obliterate the lines at the rough p. 53 cutting, but leave enough material so you can make a good finish at the line.

Cutting a Key-way.—Another job you may have frequent occasion to perform, is to cut a way for a key in a shaft and in a wheel hub. Naturally, this will be first roughed out with a cold chisel narrower than the key is to be, and also slightly shallower than the dimensions of the key.

A flat file should be used for the purpose, first a heavy rough one, for the first cutting. The better way is to have the key so it can be frequently tried while the filing process is going on, so that to fit the key in this way is a comparatively easy task.

Key-way Difficulties.—But the trouble commences when the groove is filed for the depth. Invariably, the mistake will be made of filing the width first, so the key will fit in. As a result, in deepening the groove the file will contact with the walls, and you have a key-way too wide for the key.

To avoid this, file the depth, or nearly so, and then with a fine file cut in the corners in the direction indicated by the dart, Fig. 48.

A proper key is square in cross section. In such a case the depth of the key-way, at each side wall, is just half the width of the key-way.

An excellent key-seat rule can be made by filing p. 54 out two right-angled pieces, as shown in Fig. 49, which can be attached to the ordinary six-inch metal rule, and this will enable you to scribe the line accurately for the key-way on the shaft.

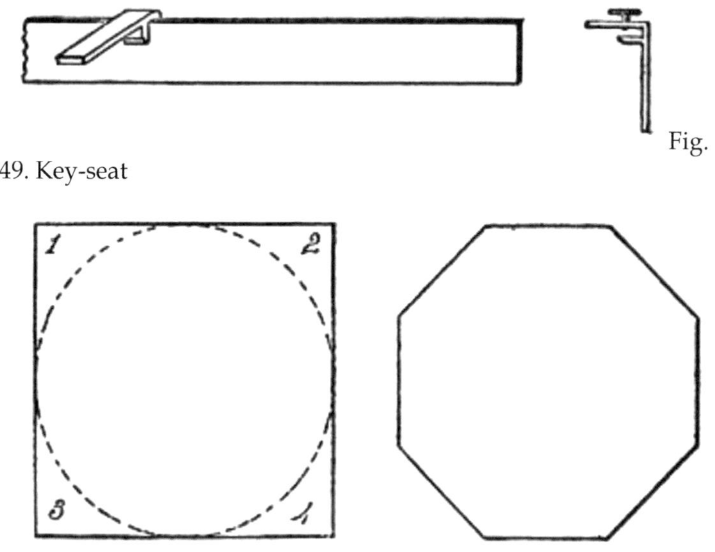

Fig. 49. Key-seat

Fig. 50.
Filing Metal Round

Fig. 51. ToList

Filing Metal Round.—It is sometimes necessary to file a piece of metal round. This is a hard job, particularly where it is impossible to scribe the end of the piece. Suppose it is necessary to file up a bearing surface, or surfaces, intermediate the ends of a square bar.

You have in that case four sides to start from, p. 55 the opposite sides being parallel with each other, so that you will have two dimensions, and four equal sides, as shown in Fig. 50.

The first step will be to file off accurately the four corners 1, 2, 3, 4, so as to form eight equal sides or faces, as shown in Fig. 51. If you will now proceed to file down carefully the eight corners, so as to make sixteen sides, as in Fig. 52, the fourth set of corners filed down will make the filed part look like the illustration Fig. 53 with thirty-two faces.

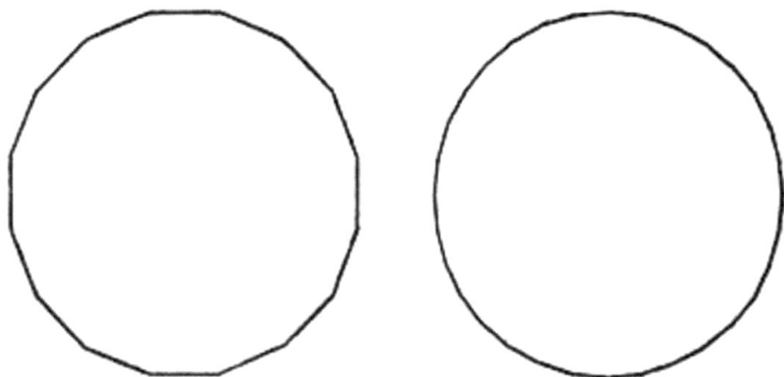

Fig. 52.   Fig. 53.
Making a Bar Round ToList

This may be further filed down into sixty-four faces, and a few cuts of the finishing file will take off the little ridges which still remain. By using emery cloth, and wrapping it around the bearing portion, and changing it continually, while drawing p. 56 it back and forth, will enable you to make a bearing which, by care, will caliper up in good shape.

Kinds of Files.—Each file has five distinct properties; namely: the length, the contour, the form in cross section, the kind of teeth, and the fineness of the teeth.

There are nine well-defined shapes for files. These may be enumerated as follows:

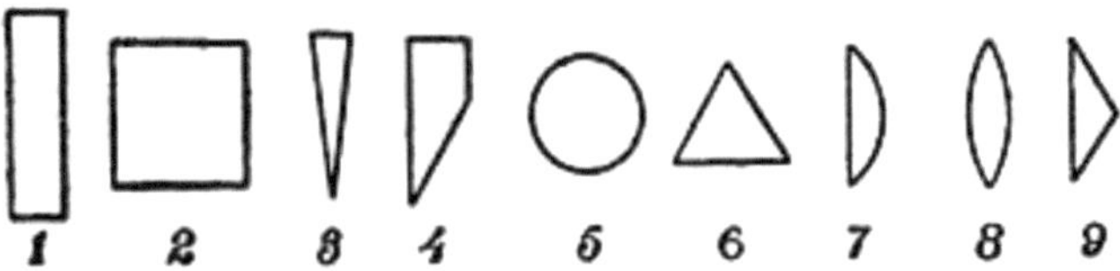

Fig. 54. Cross Sections of Files. ToList

No. 1. The cotter file. The small kind is called a verge or pivot file.

No. 2. Square file, which may be tapering from end to end, or have parallel sides throughout.

No. 3. Watch pinion file. This may have its sides parallel or tapering, to make a knife-shaped file.

No. 4. Clock-pinion; which may be used for either nicking, piecing, or squaring-off purposes.

No. 5. Round, with parallel sides for gulleting purposes, or rat-tail when it tapers.

No. 6. Triangular, or three equally-sided body for saw filing. p. 57

No. 7. Equalizing file. This is parallel when used for making clock-pinions or endless screws; or for slitting, entering, warding, or making barrel holes, when the body of the file tapers.

No. 8. Cross, or double-round, half-file.

No. 9. Slitting file; which has parallel sides only. A cant file.

Character of the File Tooth.—Files are distinguished principally by the character of the oblique, or cross grooves and ridges which do the cutting and abrading when the file is drawn across the surface.

This is really more important than the shape, because the files, by their cuttings, are adapted for the various materials which they are to be used upon.

The files are classified as *Double Cut*, of which there are the *rough, middle, bastard, second cut, smooth,* and *dead smooth.*

The *Float Cut*, which is either *rough, bastard* or *smooth;* and

The *Rasp Cut*, either *rough, bastard* or *smooth.*

Several types are illustrated in Fig. 55, which show the characteristics of the various cuts.

The rasps are used principally for soft material, such as wood or for hoofs, in horse shoeing, hence they need not be considered in connection with machine-shop work

p. 58

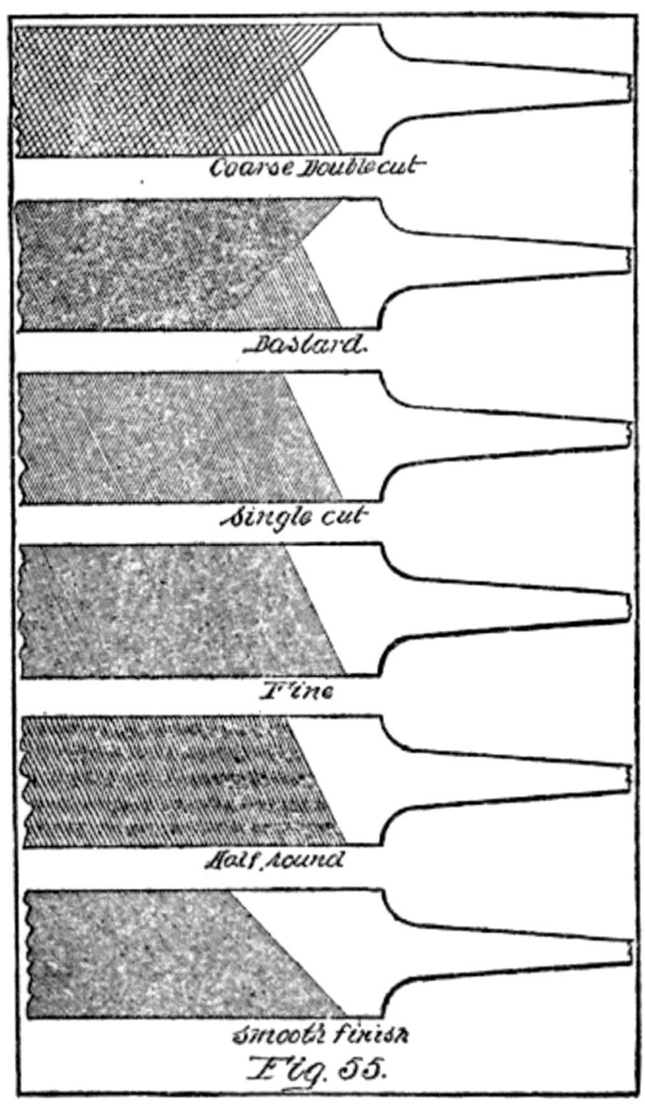

Fig. 55. Files. ToList ToList
p. 59

Holding the File.—The common mistake on the part of the begin-ner is to drag the file across the work at an angle. The body of the file should move across straight and not obliquely.

Note this movement in Fig. 56 where the dash shows the correct movement of the file with relation to the work. Also observe that the file cutting ridges are not straight across the file, but at an angle to the direction of the dart.

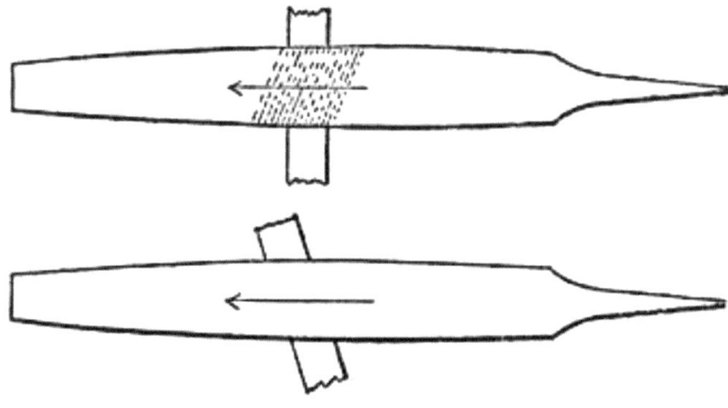

Fig. 56.

Correct File Movement ToList

Injuring Files.—Now the frequent practice is to use the file as shown in Fig. 57, in which case it is moved across obliquely. The result is that the angle of the file cut is so disposed that the teeth of the file do not properly aid in the cutting, but in a measure retard the operation.

File teeth are disposed at an angle for the pur p. 60 pose of giving them a shearing cut, which is the case when the file moves across the work on a line with its body.

To use a file as shown in Fig. 57 injures the file without giving it an opportunity to cut as fast as it would when properly used.

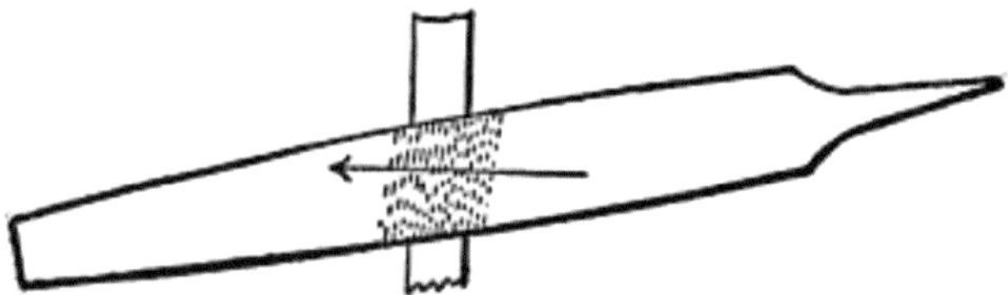

Fig. 57.

Incorrect File Movement ToList

Drawing Back the File. — In drawing back a file it is always better to allow it to drag over the work than to raise it up. It is frequently the case that some of the material will lodge in the teeth, and the back lash will serve to clear out the grooves.

This is particularly true in filing copper, aluminum, lead, and like metals, but it is well to observe this in all cases

p. 61

CHAPTER V ToC

HOW TO COMMENCE WORK

The question is often asked: Where and how shall the novice commence work?

When the shop is equipped, or partially so, sufficient, at least, to turn out simple jobs, the boy will find certain tools which are strangers to him. He must become acquainted with them and not only learn their uses, but how to use them to the best advantage.

Familiarity with Tools.—Familiarity with the appearance of tools, and seeing them in the hands of others will not be of any value. Nothing but the immediate contact with the tool will teach how to use it.

File Practice.—The file is a good tool to pick up first. Select a piece of metal, six or eight inches long, and follow the instructions laid down in the chapter relating to the use of the file.

Practice with several kinds and with different varieties of material will soon give an inkling of the best kind to use with the metal you have. Use the straight edge and the square while the filing process is going on, and apply them frequently, p. 62 to show you what speed you are making and how nearly true you are surfacing up the piece.

Using the Dividers.—Then try your hand using the dividers, in connection with a centering punch. As an example, take two pieces of metal, each about a foot long, and set the dividers to make a short span, say an inch or so, and step off the length of one piece of metal, and punch the last mark. Then do likewise with the other piece of metal, and see how nearly alike the two measurements are by comparing them.

You will find a variation in the lengths of the two measurements at the first trials, and very likely will not be able to make the two pieces register accurately after many trials, even when using the utmost care.

Sooner or later you will learn that you have not stepped paths along the two bars which were exactly straight, and this will account for the variations. In order to be accurate a line should be drawn along each piece of metal, and the dividers should step off the marks on that line.

Finding Centers.—By way of further experiment, it might be well to find the exact center of the ends of a square bar, putting in the punch marks and then mounting it in the lathe centers to see how accurately this has been done.

If either end is out of true the punch marks can p. 63 be corrected by inclining the punch, so that when it is struck it will move over the point in the direction of its true center. This may be followed up by centering the end of a round bar so as to make it true. This will be found to be a more difficult job, unless you have a center head, a tool made for that purpose.

It is good practice, however, to make trials of all this work, as it will enable you to judge of measurements. It can be done with the dividers by using care in scribing the centers.

Hack-Saw Practice.—Practice with the hack-saw should be indulged in frequently. Learn to make a straight cut through a bar. Try to do this without using a square to guide you. One of the tests of a good mechanic is ability to judge a straight cut.

The following plan is suggested as a test for the eye. Use a bar of iron or steel one inch square, and make a cut an eighth of an inch deep across it; then turn it around a quarter, so as to expose the nest face, and continue the cut along the side, the same depth, and follow this up with the remaining two sides, and see how near the end of the first cut and the finish cut come together. The test will surprise you.

Cutting Metals True.—When you saw off the end of such a bar for trial purposes, use a square, p. 64 after the cut is made, and note how much it is out of true in both directions. It is a curious fact that most mechanics are disposed to saw or cut crooked in one direction, either to the right or to the left. In tests made it is found that this defect is persisted in.

It is practice only which will remedy this, and it would be well for the boy to learn this for himself as early in his career as possible, and correct the tendency to veer in either direction.

The test of sawing around a round bar is also commended. After a few trials you will be surprised to see how your judgment will improve in practice.

Lathe Work.—Learn the uses of the chuck. As you have, probably, economized as much as possible, a universal chuck is not available, hence the first experience will be with an independent chuck, where the three dogs move independently of each other. This will give you some work to learn how you can get the job true.

Now, before attempting to cut the material, thoroughly learn all the parts of the feed mechanism, and how to reverse, as well as to cross feed. Learn the operation of the operative parts so that your hand will instinctively find them, while the eye is on the work.

First Steps.—See to it that your tools are sharp, p. 65 and at the first trials make light cuts. Practice the feeds by manually moving the tool holder, for surface cutting as well as for cross cutting.

Setting the Tool.—Set the cutting tool at various angles, and try the different tools, noting the peculiarities of each, at the different speeds. Do not, by any means, use refractory metals for your first attempt. Mild steel is a good test, and a light gray iron is admirable for practice lessons.

Metals Used.—Brass is good for testing purposes, but the difficulty is that the tendency of the boy, at first, is to try to do the work too rapidly, and brass encourages this tendency. Feed slowly and regularly until you can make an even finish.

Then chuck and re-chuck to familiarize yourself with every operative part of the lathe, and never try to force the cutting tool. If it has a tendency to run into the work, set it higher. If, on the other hand, you find, in feeding, that it is hard to move the tool post along, the tool is too high, and should be lowered.

The Four Important Things.—Constant practice of this kind will soon enable you to feel instinctively when the tool is doing good work. While you are thus experimenting do not forget the speed. This will need your attention.

Remember, you have several things to think about in commencing to run the lathe, all of which p. 66 will take care of themselves when it becomes familiar to you. These may be enumerated as follows:

First: The kind of tool best to use.

Second: Its proper set, to do the best work.

Third: The speed of the work in the lathe.

Fourth: The feed, or the thickness of the cut into the material.

Turning up a Cylinder.—The first and most important work is to turn up a small cylinder to a calipered dimension. When it is roughed down ready for the finish cut, set the tool so it will take off a sufficient amount to prevent the caliper from spanning it, and this will enable you to finish it off with emery paper, or allow another small cut to be taken.

Turning Grooves.—Then follow this up by turning in a variety of annular grooves of different depths and widths; and also V-shaped grooves, the latter to be performed by using both the longitudinal and transverse feeds. This will give you excellent practice in using both hands simultaneously.

The next step would be to turn out a bore and fit a mandrel into it. This will give you the opportunity to use the caliper to good advantage, and will test your capacity to use them for inside as well as for outside work

p. 67

Discs.—A job that will also afford good exercise is to turn up a disc with a groove in its face, and then chuck and turn another disk with an annular rib on its face to fit into the groove. This requires delicacy of measurement with the inside as well as the outside calipers.

The groove should be cut first, and the measurement taken from that, as it is less difficult to handle and set the tool for the rib than for the groove.

Lathe Speeds.—Do not make the too common mistake of running the mandrel at high speeds in your initial tests. It is far better to use a slow speed, and take a heavy cut. This is good advice at all times, but it is particularly important with beginners

p. 68

CHAPTER VI ToC

ILLUSTRATING SOME OF THE FUNDAMENTAL DEVICES

There are numerous little devices and shop expedients which are desirable, and for which the boy will find uses as he progresses.

We devote this chapter to hints of this kind, all of which are capable of being turned out or utilized at various stages.

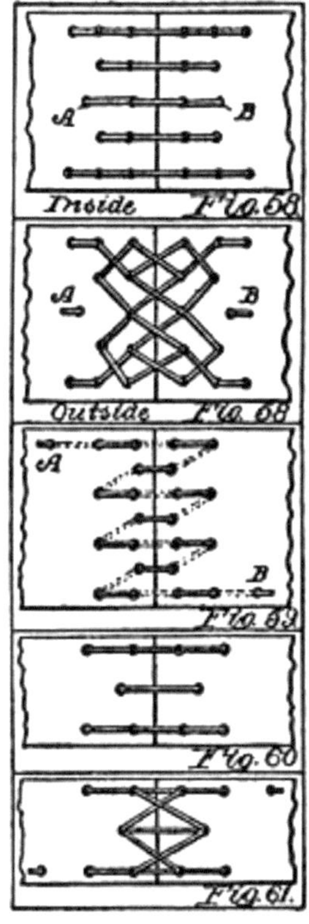

ToList

Lacing Belts. — To properly lace a belt is quite an art, as many who have tried it know. If a belt runs off the pulley it is attributable to one of three causes: either the pulleys are out of line or the shafts are not parallel or the belt is laced so it makes the belt longer at one margin than the other.

In Fig. 58 the lacing should commence at the center hole (A) of one belt end and lace outwardly, terminating at the hole (B) in the center of the other belt end, as shown in Fig. 58.

In Fig. 59 the lacing commences at A, and terminates at the hole (B) at the edge. This will be ample for all but the widest belts.

Fig. 60 is adapted for a narrow belt. The lacing commences at one margin hole (A), and terminates at the other margin hole (Z)

p. 69

Fig. 61 shows the outside of the belt.

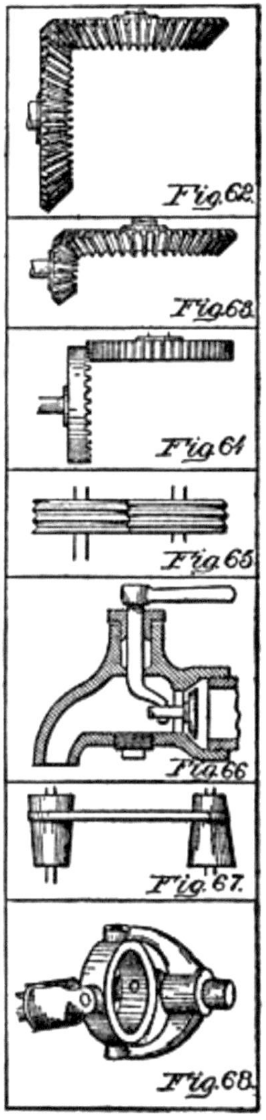

ToList

Fig. 62. Gears.—This is something every boy ought to know about. Fig. 62 shows a pair of intermeshing bevel gears. This is the correct term for a pair when both are of the same diameter.

Miter Gears.—In Fig. 63 we have a pair of miter gears, one being larger than the other. Remember this distinction.

Fig. 64. Crown Wheel.—This is a simple manner of transmitting motion from one shaft to another, when the shafts are at right angles, or nearly so, without using bevel or miter gears.

Fig. 65. Grooved Friction Gearing.—Two grooved pulleys, which fit each other accurately, will transmit power without losing too much by friction. The deeper the grooves the greater is the loss by friction.

Fig. 66. A Valve Which Closes by the Water Pressure.—The bibb has therein a movable valve on a horizontal stem, the valve being on the inside p. 70 of the seat. The stem of the handle has at its lower end a crank bend, which engages with the outer end of the valve stem. When the handle is turned in either direction the valve is unseated. On releasing the handle the pressure of the water against the valve seats it.

Fig. 67. Cone Pulleys.—Two cone pulleys of equal size and taper provide a means whereby a change in speed can be transmitted from one shaft to another by merely moving the belt to and fro. The slightest change is available by this means.

Fig. 68. Universal Joint.—A wheel, with four projecting pins, is placed between the U-shaped yokes on the ends of the approaching shafts. The pins serve as the pivots for the angles formed by the two shafts.

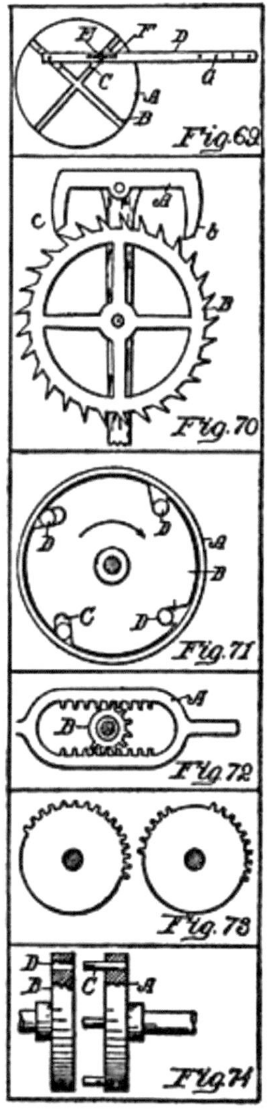

ToList

Fig. 69. Trammel for Making an Ellipse.—This is a tool easily made, which will be of great service in the shop. In a disc (A), preferably made of brass, are two channels (B) at right angles to each other. The grooves are undercut, so that the blocks (C) will fit and slide in the grooves and be held therein by the dove-tailed formation. Each block is longer than the width of the groove, and has an outwardly projecting pin which passes through a bar (D). One pin (E) is movable along in a slot, but is adjustable at any point so that the shape of the ellipse may be p. 71 varied. The end of the bar has a series of holes (G) for a pencil, so that the size of the ellipse may also be changed.

Fig. 70. Escapements.—Various forms of escapements may be made, but the object of all is the same. The device is designed to permit a wheel to move intermittingly or in a step by step movement, by the swinging motion of a pendulum. Another thing is accomplished by it. The teeth of the escapement are cut at such an angle that, as one of the teeth of the escapement is released from one tooth of the escapement wheel, the spring, or the weight of the clock, will cause one of the teeth of the escapement wheel to engage the other tooth of the escapement, and give the pendulum an impulse in the other direction. In the figure, A is the escapement, B the escapement wheels and *a*, *b*, the p. 72 pallets, which are cut at suitable angles to actuate the pendulum.

Fig. 71. Simple Device to Prevent a Wheel or Shaft prom Turning Back.—This is a substitute for a pawl and ratchet wheel. A is a drum or a hollow wheel and B a pulley on a shaft, and this pulley turns loosely with the drum (A). Four tangential slots (C) are cut into the perimeter of the pulley (B), and in each is a hardened steel roller (D). It matters not in what position the wheel (B) may be, at least two of the rollers will always be in contact with the inside of the drum (A), and thus cause the pulley and drum to turn together. On reversing the direction of the pulley the rollers are immediately freed from binding contact.

Fig. 72. Racks and Pinions.—The object of this form of mechanism is to provide a reciprocating, or back-and-forth motion, from a shaft which turns continually in one direction. A is the rack and B a mutilated gear. When the gear turns it moves the rack in one direction, because the teeth of the gear engage the lower rack teeth, and when the rack has moved to the end its teeth engage the teeth of the upper rack, thus reversing the movement of the rack.

Fig. 73. Mutilated Gears.—These are made in so many forms, and adapted for such a p. 73 variety of purposes, that we merely give a few samples to show what is meant by the term.

Fig. 74. Simple Shaft Coupling.—Prepare two similarly formed discs (A, B), which are provided with hubs so they may be keyed to the ends of the respective shafts. One disc has four or more projecting pins (C), and the other disc suitable holes (D) to receive the pins.

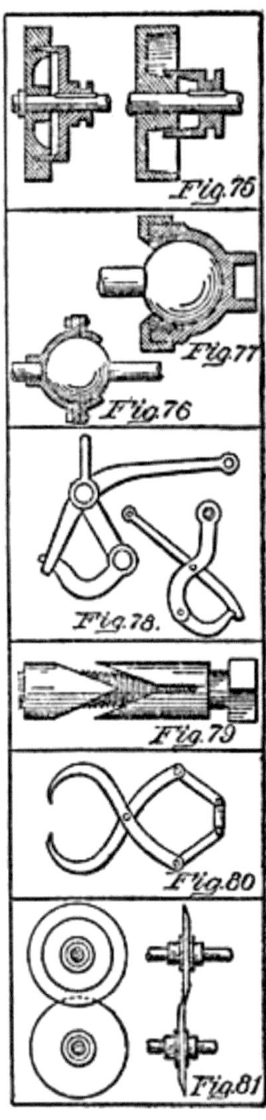

ToList

Fig. 75. Clutches.—This is a piece of mechanism which is required in so many kinds of machinery, that we show several of the most approved types.

Fig. 76. Ball and Socket Joints.—The most practical form of ball and socket joints is simply a head in which is a bowl-shaped cavity the depth of one-half of the ball. A plate with a central opening small enough to hold in the ball, and p. 74 still large enough at the neck to permit the arm carrying the ball to swing a limited distance, is secured by threads, or by bolts, to the head. The first figure shows this.

Fig. 77 illustrates a simple manner of tightening the ball so as to hold the standard in any desired position.

Fig. 78. Tripping Devices.—These are usually in the form of hooks, so arranged that a slight pull on the tripping lever will cause the suspended articles to drop.

Fig. 79. Anchor Bolt.—These are used in brick or cement walls. The bolt itself screws into a sleeve which is split, and draws a wedge nut up to the split end of the sleeve. As a result the split sleeve opens or spreads out and binds against the wall sufficiently to prevent the bolt from being withdrawn.

Fig. 80. Lazy Tongs.—One of the simplest and most effective instruments for carrying ice, boxes or heavy objects, which are bulky or inconvenient to carry. It grasps the article firmly, and the heavier the weight the tighter is its grasp.

Fig. 81. Disc Shears.—This is a useful tool either for cutting tin or paper, pasteboard and the like. It will cut by the act of drawing the material through it, but if power is applied to one or to both of the shafts the work is much facilitated, p. 75 particularly in thick or hard material.

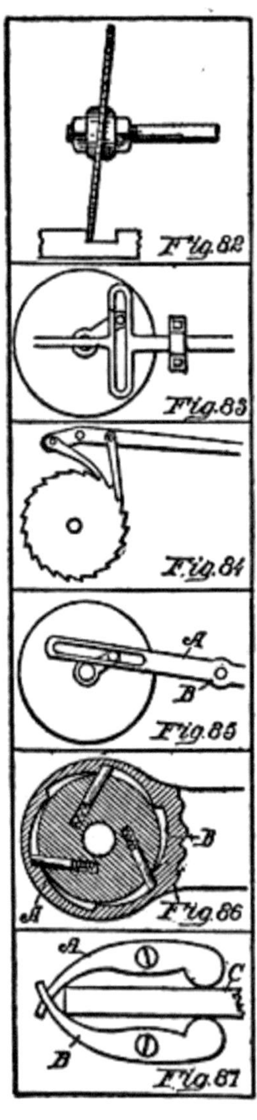

ToList

Fig. 82. Wabble Saw.—This is a most simple and useful tool, as it will readily and quickly saw out a groove so that it is undercut. The saw is put on the mandrel at an angle, as will be seen, and should be run at a high rate of speed.

Fig. 83. Crank Motion by a Slotted Yoke.—This produces a straight back-and-forth movement from the circular motion of a wheel or crank. It entirely dispenses with a pitman rod, and it enables the machine, or the part of the machine operated, to be placed close to the crank.

Fig. 84. Continuous Feed by the Motion of a Lever.—The simple lever with a pawl on each side of the fulcrum is the most effective means to make a continuous feed by the simple movement of a lever. The form shown is capable of p. 76 many modifications, and it can be easily adapted for any particular work desired.

Fig. 85. Crank Motion.—By the structure shown, namely, a slotted lever (A), a quick return can be made with the lever. B indicates the fulcrum.

Fig. 86. Ratchet Head.—This shows a well-known form for common ratchet. It has the advantage that the radially movable plugs (A) are tangentially disposed, and rest against walls (B) eccentrically disposed, and are, therefore, in such a position that they easily slide over the inclines.

Fig. 87. Bench Clamp.—A pair of dogs (A, B), with the ends bent toward each other, and pivoted midway between the ends to the bench in such a position that the board (C), to be held between them, on striking the rear ends of the dogs, will force the forward ends together, and thus clamp p. 77 it firmly for planing or other purposes.

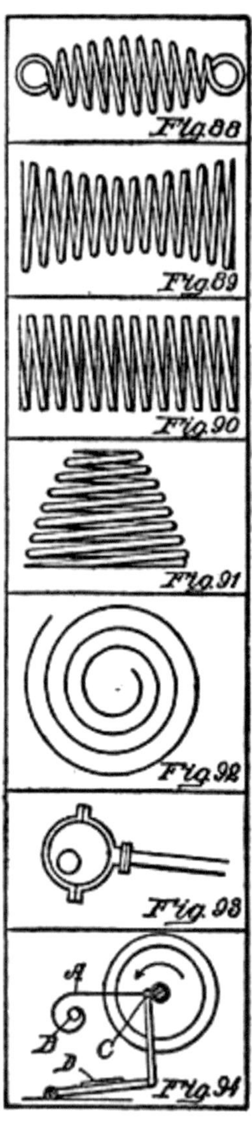

ToList

Fig. 88. Helico-Volute Spring.—This is a form of spring for tension purposes. The enlarged cross-section of the coil in its middle portion, with the ends tapering down to the eyes, provides a means whereby the pull is transferred from the smaller to the larger portions, without producing a great breaking strain near the ends.

Fig. 89. Double Helico-Volute.—This form, so far as the outlines are considered, is the opposite of Fig. 88. A compression spring of this kind has a very wide range of movement.

Fig. 90. Helical Spring.—This form of coil, uniform from end to end, is usually made of metal which is square in cross-section, and used where it is required for heavy purposes

p. 78

Fig. 91. Single Volute Helix-Spring.—This is also used for compression, intended where tremendous weights or resistances are to be overcome, and when the range of movement is small.

Fig. 92. Flat Spiral, or Convolute.—This is for small machines. It is the familiar form used in watches owing to its delicate structure, and it is admirably adapted to yield to the rocking motion of an arbor.

Fig. 93. Eccentric Rod and Strap.—A simple and convenient form of structure, intended to furnish a reciprocating motion where a crank is not available. An illustration of its use is shown on certain types of steam engine to operate the valves.

Fig. 94. Anti-Dead Center for Foot-Lathes.—A flat, spiral spring (A), with its coiled end attached to firm support (B), has its other end pivotally attached to the crank-pin (C), the tension of the spring being such that when the lathe stops the crack-pin will always be at one side of the dead-center, thus enabling the operator to start the machine by merely pressing the foot downwardly on the treadle (D)

p. 79

CHAPTER VII ToC

PROPERTIES OF MATERIALS

A workman is able to select the right metals because he knows that each has some peculiar property which is best adapted for his particular use. These with their meaning will now be explained.

Elasticity. — This exists in metals in three distinct ways: First, in the form of *traction*. Hang a weight on a wire and it will stretch a certain amount. When the weight is removed the wire shrinks back to its original length.

Second: If the weight on the wire is rotated, so as to twist it, and the hand is taken from the weight, it will untwist itself, and go back to its original position. This is called *torsion*.

Third: A piece of metal may be coiled up like a watch spring, or bent like a carriage spring, and it will yield when pressure is applied. This is called *flexure*.

Certain kinds of steel have these qualities in a high degree.

Tenacity. — This is a term used to express the resistance which the body opposes to the separation of its parts. It is determined by forming p. 80 the metal into a wire, and hanging on weights, to find how much will be required to break it. If we have two wires, the first with a transverse area only one-quarter that of the second, and the first breaks at 25 pounds, while the second breaks at 50 pounds, the tenacity of the first is twice as great as that of the second.

To the boy who understands simple ratio in mathematics, the problem would be like this:

$25 \times 4 : 50 \times 1$, or as $2 : 1$.

The Most Tenacious Metal. — Steel has the greatest tenacity of all metals, and lead the least. In proportion to weight, however, there are many substances which have this property in a higher degree. Cotton fibers will support millions of times their own weight.

There is one peculiar thing, that tenacity varies with the form of the body. A solid cylindrical body has a greater strength than a square one of the same size; and a hollow cylinder more tenacity than a solid one. This principle is well known in the bones of animals, in the feathers of birds, and in the stems of many plants.

In almost every metal tenacity diminishes as the temperature increases.

Ductility. — This is a property whereby a metal may be drawn out to form a wire. Some metals, p. 81 like cast iron, have absolutely no ductility. The metal which possesses this property to the highest degree, is platinum. Wires of this metal have been drawn out so fine that over 30,000 of them laid side by side would measure only one inch across, and a mile of such wire would weigh only a grain, or one seven-thousandth of a pound.

Malleability. — This is considered a modification of ductility. Any metal which can be beaten out, as with a hammer, or flattened into sheets with rollers, is considered malleable. Gold possesses this property to the highest degree. It has been beaten into leaves one three-hundred-thousandth of an inch thick.

Hardness. — This is the resistance which bodies offer to being scratched by others. As an example, the diamond has the capacity to scratch all, but cannot be scratched by any other.

Alloys. — Alloys, that is a combination of two or more metals, are harder than the pure metals, and for this reason jewelry, and coins, are usually alloyed.

The resistance of a body to compression does not depend upon its hardness. Strike a diamond with a hammer and it flies to pieces, but wood does not. One is brittle and the other is tough.

The machinist can utilize this property by understanding that velocity enables a soft material p. 82 to cut a harder one. Thus, a wrought iron disc rotating rapidly, will cut such hard substances as agate or quartz.

Resistance. — All metals offer more or less resistance to the flow of an electric current. Silver offers the least resistance, and German silver the greatest. Temperature also affects the flow. It passes more easily over a cold than a warm conductor.

Persistence. — All metals on receiving heat, will retain it for a certain length of time, and will finally cool down to the temperature of the surrounding atmosphere. Some, like aluminum, retain it for a long time; others, as iron, will give it off quickly.

Conductivity.—All metals will conduct heat and cold, as well as electricity. If one end of a metal bar is heated, the heat creeps along to the other end until it has the same temperature throughout. This is called *equalization.*

If a heated bar is placed in contact with another, the effect is to increase the temperature of the cold bar and lower that of the warm bar. This is called *reciprocity.*

Molecular Forces.—*Molecular* attraction is a force which acts in such a way as to bring all the particles of a body together. It acts in three p. 83 ways, dependent on the particular conditions which exist.

First: *Cohesion.* This exists between molecules which are of the same kind, as for instance, iron. Cohesion of the particles is very strong in solids, much weaker in liquids, and scarcely exists at all between the particles in gases.

Second: *Adhesion* is that property which exists between the surfaces of bodies in contact. If two flat surfaces are pressed together, as for instance, two perfectly smooth and flat pieces of lead, they will adhere. If, for instance, oil should be put on the surfaces, before putting them together, they would adhere so firmly that it would be difficult to pull them apart.

Third: *Affinity.* This is another peculiarity about materials. Thus, while cohesion binds together the molecules of water, it is chemical affinity which unites two elements, like hydrogen and oxygen, of which water is composed.

Porosity.—All matter has little hollows or spaces between the molecules. You know what this is in the case of a sponge, or pumice stone. Certain metals have the pores so small that it is difficult to see them except with a very powerful glass. Under great pressure water can be forced through the pores of metals, as has been done in p. 84 the case of gold. Water also is porous, but the spaces between the molecules are very small.

Compressibility.—It follows from the foregoing statement, that if there are little interstices between the molecules, the various bodies can be compressed together. This can be done in varying degrees with all solids, but liquids, generally, have little compressibility. Gases are readily reduced in volume by compression.

Elasticity.—This is a property by virtue of which a body resumes its original form when compressed. India rubber, ivory and glass are examples of elasticity; whereas, lead and clay do not possess this property. Air is the most elastic of all substances.

Inertia.—This is a property of matter by virtue of which it cannot of itself change its state of motion or of rest.

Newton's first law of motion is, in substance, that matter at rest will eternally remain at rest, and matter in motion will forever continue in motion, unless acted on by some external force.

A rider is carried over the head of a horse when the latter suddenly stops. This illustrates the inertia of movement. A stone at rest will always remain in that condition unless moved by some force. That shows the inertia of rest.

Momentum.—This is the term to designate the p. 85 quantity of motion in a body. This quantity varies and is dependent on the mass, together with the velocity. A fly wheel is a good example. It continues to move after the impelling force ceases; and a metal wheel has greater momentum than a wooden wheel at the same speed, owing to its greater mass.

If, however, the wooden wheel is speeded up sufficiently it may have the same momentum as the metal one.

Weight.—All substances have what is called *weight*. This means that everything is attracted toward the earth by the force of gravity. Gravity, however, is different from weight. All substances attract each other; not only in the direction of the center of the earth, but laterally, as well.

Weight, therefore, has reference to the pull of an object toward the earth; and gravity to that influence which all matter has for each other independently of the direction.

Centripetal Force.—This attraction of the earth, which gives articles the property of weight, is termed centripetal force—that is, the drawing in of a body.

Centrifugal Force.—The direct opposite of centripetal, is centrifugal force, which tends to p. 86 throw outwardly. Dirt flying from a rapidly moving wheel illustrates this.

Capillary Attraction.—There is a peculiar property in liquids, which deserves attention, and should be understood, and that is the name given to the tendency of liquids to rise in fine tubes.

It is stated that water will always find its level. While this is true, we have an instance where, owing to the presence of a solid, made in a peculiar form, causes the liquid, within, to rise up far beyond the level of the water.

This may be illustrated by three tubes of different internal diameters. The liquid rises up higher in the second than in the first, and still higher in the third than in the second. The smaller the tube the greater the height of the liquid.

This is called *capillary attraction*, the word capillary meaning a hair. The phenomena is best observed when seen in tubes which are as fine as hairs. The liquid has an affinity for the metal, and creeps up the inside, and the distance it will thus move depends on the size of the tube.

The Sap of Trees.—The sap of trees goes upwardly, not because the tree is alive, but due to this property in the contact of liquids with a solid. It is exactly on the same principle that if the end of a piece of blotting paper is immersed p. 87 in water, the latter will creep up and spread over the entire surface of the sheet.

In like manner, oil moves upwardly in a wick, and will keep on doing so, until the lighted wick is extinguished, when the flow ceases. When it is again lighted the oil again flows, as before.

If it were not for this principle of capillary attraction, it would be difficult to form a bubble of air in a spirit level. You can readily see how the liquid at each end of the air bubble rounds it off, as though it tried to surround it.

Sound.—Sound is caused by vibration, and it would be impossible to convey it without an elastic medium of some kind.

Acoustics is a branch of physics which treats of sounds. It is distinguished from music which has reference to the particular kinds.

Sounds are distinguished from *noises*. The latter are discordant and abrupt vibrations, whereas the former are regular and continuous.

Sound Mediums.—- Gases, vapors, liquids and solids transmit vibrations, but liquids and solids propagate with greater velocity than gases.

Vibration.—A vibration is the moving to and fro of the molecules in a body, and the greater their movement the more intense is the sound. The intensity of the sound is affected by the density of the atmosphere, and the movement p. 88 of the winds also changes its power of transmission.

Sound is also made more intense if a sonorous body is near its source. This is taken advantage of in musical instruments, where a sounding-board is used, as in the case of the piano, and in the violin, which has a thin shell as a body for holding the strings.

Another curious thing is shown in the speaking tube, where the sound waves are confined, so that they are carried along in one line, and as they are not interfered with will transmit the vibrations to great distances.

Velocity of Sound.—The temperature of the air has also an effect on the rate of transmission, but for general purposes a temperature of 62 degrees has been taken as the standard. The movement is shown to be about 50 miles in 4 minutes, or at the rate of 1,120 feet per second.

In water, however, the speed is four times greater; and in iron nearly fifteen times greater. Soft earth is a poor conductor, while rock and solid earth convey very readily. Placing the ear on a railway track will give the vibrations of a moving train miles before it can be heard through the air.

Sound Reflections.—Sound waves move outwardly from the object in the form of wave-like p. 89 rings, but those concentric rings, as they are called, may be interrupted at various points by obstacles. When that is the case the sound is buffeted back, producing what is called echoes.

Resonance.—Materials have a quality that produces a very useful result, called *resonance*, and it is one of the things that gives added effect to a speaker's voice in a hall, where there is a constant succession of echoes. A wall distant from the speaker about 55 feet, produces an almost instantaneous reflection of the sound, and at double that measurement the effect is still stronger. When the distance is too short for the reflecting sound to be heard, we have *resonance*. It enriches the sound of the voice, and gives a finer quality to musical instruments.

Echoes.—When sounds are heard after the originals are emitted they tend to confusion, and the quality of resonance is lost. There are places where echoes are repeated many times. In the chateau of Simonetta, Italy, a sound will be repeated thirty times.

Speaking Trumpet.—This instrument is an example of the use of reflection. It is merely a bell-shaped, or flaring body, the large end of which is directed to the audience. The voice talking into the small end is directed forwardly, and is reflected from the sides, and its resonance also enables the p. 90 vibrations to carry farther than without the use of the solid part of the instrument.

The ear trumpet is an illustration of a sound-collecting device, the waves being brought together by reflection.

The Stethoscope.—This is an instrument used by physicians, and it is so delicate that the movements of the organs of the body can be heard with great distinctness. It merely collects the vibrations, and transmits them to the ears by the small tubes which are connected with the collecting bell.

The Vitascope.—- Numerous instruments have been devised to determine the rate of vibration of different materials and structures, the most important being the *vitascope*, which has a revolvable cylinder, blackened with soot, and this being rotated at a certain speed, the stylus, which is attached to the vibrating body, in contact with the cylinder, will show the number per second, as well as the particular character of each oscillation.

The Phonautograph.—This instrument is used to register the vibration of wind instruments, as well as the human voice, and the particular forms of the vibrations are traced on a cylinder, the tracing stylus being attached to a thin vibrating membrane which is affected by the voice or instrument.

The Phonograph.—This instrument is the outgrowth of the stylus forms of the apparatus de p. 91 scribed, but in this case the stylus, or needle, is fixed to a metallic diaphragm, and its point makes an impression on suitable material placed on the outside of a revolvable cylinder or disc.

Light.-Light is the agent which excites the sensation of vision in the eye. Various theories have been advanced by scientists to account for the phenomenon, and the two most noted views are the *corpuscular*, promulgated by Sir Isaac Newton, and the *undulatory*, enunciated by Huygens and Euler.

The *corpuscular* theory conceives that light is a substance of exceedingly light particles which are shot forth with immense velocity. The *undulatory* theory, now generally accepted, maintains that light is carried by vibrations in ether. Ether is a subtle elastic medium which fills all space.

Luminous bodies are those like the sun, which emit light. Rays may *diverge*, that is, spread out; *converge*, or point toward each other; or they may be *parallel* with each other.

Velocity of Light.—Light moves at the rate of about 186,000 miles a second. As the sun is about 94,000,000 miles from the earth, it takes 8 1/2 minutes for the light of the sun to reach us.

Reflection.—One of the most important things connected with light is that of reflection. It is that quality which is utilized in telescopes, micro p. 92 scopes, mirrors, heliograph signaling and other like apparatus and uses. The underlying principle is, that a ray is reflected, or thrown back from a mirror at the same angle as that which produces the light.

When the rays of the sun, which are, of course, parallel, strike a concave mirror, the reflecting rays are converged; and when the rays strike a convex mirror they diverge. In this way the principle is employed in reflecting telescopes.

Refraction.—This is the peculiar action of light in passing through substances. If a ray passes through water at an angle to the surface the ray will bend downwardly in passing through, and then again pass on in a straight line. This will be noticed if a pencil is stood in a glass of water at an angle, when it will appear bent.

Refraction is that which enables light to be divided up, or analyzed. In this way white light from the sun is shown to be composed of seven principal colors.

Colors.—If the light is passed through a prism, which is a triangularly shaped piece of glass, the rays on emerging will diverge from each other, and when they fall on a wall or screen the colors red, orange, yellow, green, blue, indigo and violet are shown

p. 93

The reason for this is that the ray in passing through the prism has the different colors in it refract at different angles, the violet bending more than the red.

The Spectroscope.—The ability to make what is thus called a *spectrum*, brought forth one of the most wonderful instruments ever devised by man. If any metal, or material, is fused, or put in such a condition that a ray of light can be obtained from it, and this light is passed through a prism, it will be found that each substance has its own peculiar divisions and arrangements of colors.

In this way substances are determined by what is called *spectrum analysis*, and it is by means of this instrument that the composition of the sun, and the planets and fixed stars are determined.

The Rainbow.—The rainbow is one of the effects of refraction, as the light, striking the little globular particles of water suspended in the air, produces a breaking up of the white light into its component colors, and the sky serves as a background for viewing the analysis thus made.

Heat.—It is now conclusively proven, that heat, like light, magnetism and electricity, is merely a mode of motion.

The *mechanical* theory of heat may be shown by rubbing together several bodies. Heat expands p. 94 all substances, except ice, and in expanding develops an enormous force.

Expansion. — In like manner liquids expand with heat. The power of mercury in expanding may be understood when it is stated that a pressure of 10,000 pounds would be required to prevent the expansion of mercury, when heated simply 10 degrees.

Gases also expand. While water, and the different solids, all have their particular units of expansion, it is not so with gases, as all have the same coefficient

p. 95

CHAPTER VIII ToC

HOW DRAUGHTING BECOMES A VALUABLE AID

The ability to read drawings is a necessary part of the boy's education. To know how to use the tools, is still more important. In conveying an idea about a piece of mechanism, a sketch is given. Now, the sketch may be readable in itself, requiring no explanation, or it may be of such a nature that it will necessitate some written description.

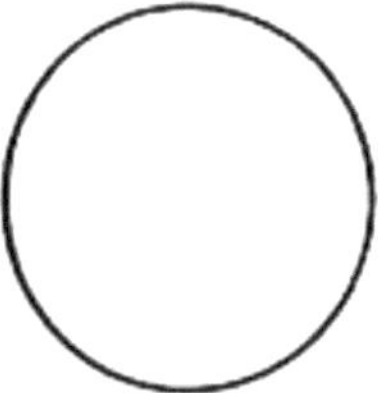

Fig. 95. Plain Circle ToList

Lines in Drawing. —- In drawing, lines have a definite meaning. A plain circular line, like Fig. 95, when drawn in that way, conveys three meanings: It may represent a rim, or a bent piece of wire; it may illustrate a disk; or, it may convey the idea of a ball.

Suppose we develop them to express the three forms accurately. Fig. 96, by merely adding an p. 96 interior line, shows that it is a rim. There can be no further doubt about that expression.

Fig. 97 shows a single line, but it will now be noticed that the line is thickened at the lower right-hand side, and from this you can readily infer that it is a disk.

Shading. —Fig. 98, by having a few shaded lines on the right and lower side, makes it have the appearance of a globe or a convex surface.

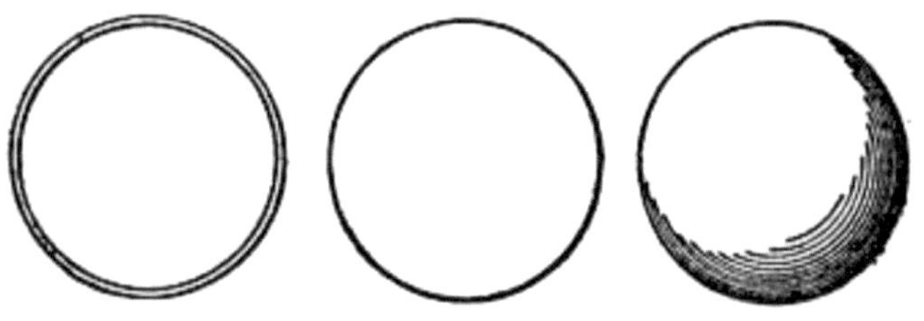

Ring - Raised Surface - Sphere ToList

Shading or thickening the lines also gives another expression to the same circular line.

In Fig. 99, if the upper and left-hand side of the circle is heavily shaded, it shows that the area within the circle is depressed, instead of being raised.

Direction of Shade.—On the other hand, if the shading lines, as in Fig. 100, are at the upper left-hand side, then the mind at once grasps the idea of a concave surface.

The first thing, therefore, to keep in mind, is this p. 97 fact: That in all mechanical drawing, the light is supposed to shine down from the upper left-hand corner and that, as a result, the lower vertical line, as well as the extreme right-hand vertical line, casts the shadows, and should, therefore, be made heavier than the upper horizontal, and the left-hand vertical lines.

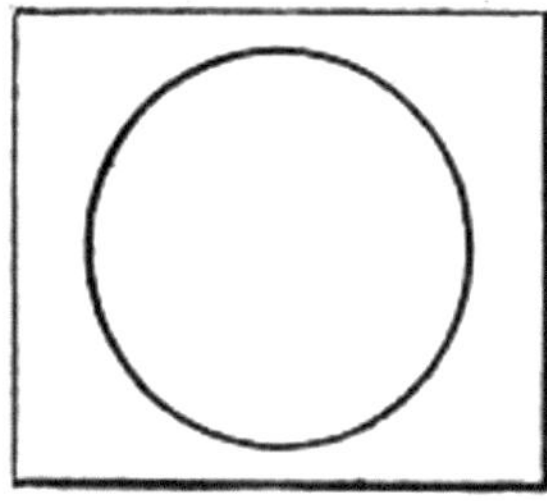 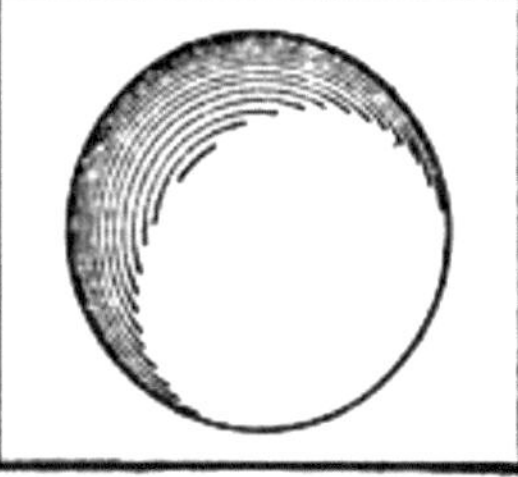

Depressed Surface - Concave ToList

There are exceptions to this rule, which will be readily understood by following out the illustrations in the order given below.

Perspectives. — The utility of the heavy lines will be more apparent when drawing square, rectangular, or triangular objects.

Let us take Fig. 101, which appears to be the perspective of a cube. Notice that all lines are of the same thickness. When the sketch was first brought to me I thought it was a cube; but the explanation which followed, showed that the man who p. 98 made the sketch had an entirely different meaning.

He had intended to convey to my mind the idea of three pieces, A, B, C, of metal, of equal size, joined together so as to form a triangularly shaped pocket as shown in Fig. 101. The addition of the inner lines, like D, quickly dispelled the suggestion of the cube.

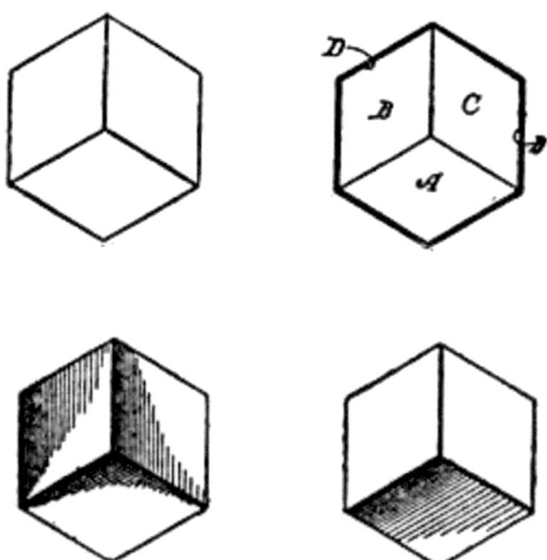

Figs. 101-104. Forms of Cubical Outlines ToList

"But," he remarked, "I want to use the thinnest metal, like sheets of tin; and you show them thick by adding the inner lines."

Such being the case, if we did not want to show p. 99 thickness as its structural form, we had to do it by making the lines themselves and the shading give that structural idea. This was done by using the single lines, as in Fig. 103, and by a slight shading of the pieces A, B, C.

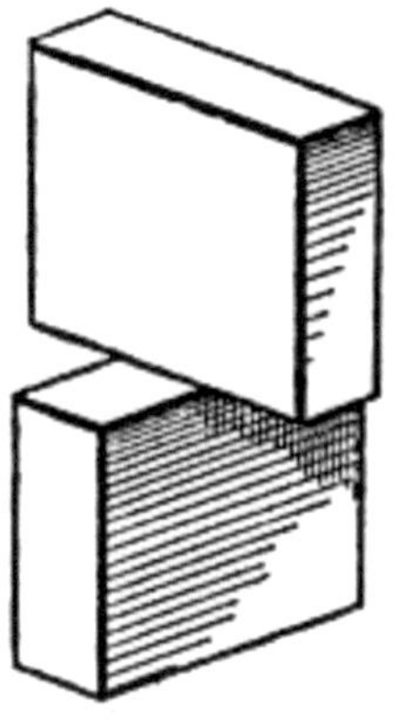

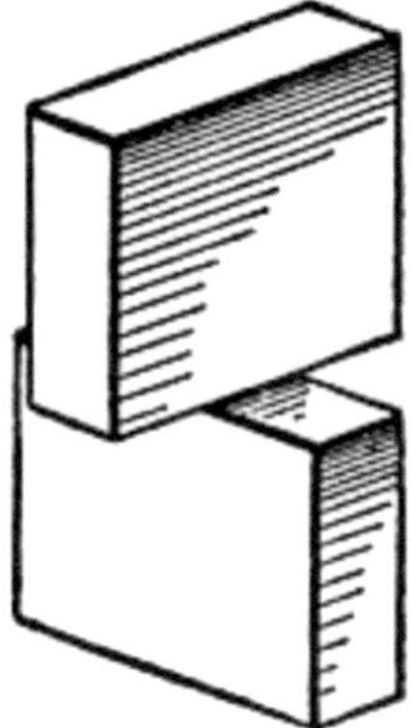

Fig. 105.   Fig. 106.

Shading Edges ToList

The Most Pronounced Lines.—If it had been a cube, or a solid block, the corners nearest the eye would have been most pronounced, as in Fig. 104, and the side next to the observer would have been darkest.

This question of light and shadow is what expresses the surface formation of every drawing. Simple strokes form outlines of the object, but their thickness, and the shading, show the character enclosed by the lines. p. 100 Direction of Light.—Now, as stated, the casting of the shadow downward from the upper left-hand corner makes the last line over which it passes the thickest, and in Figs. 105 and 106 they are not the extreme lines at the bottom and at the right side, because of the close parallel lines.

In Figs. 109 and 110 the blades superposed on the other are very thin, and the result is the lines at the right side and bottom are made much heavier.

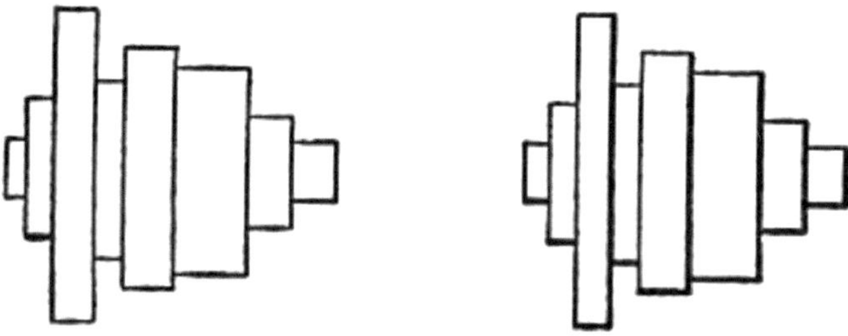

Fig. 107.   Fig. 108.

Illustrating Heavy Lines ToList

This is more fully shown in Figs. 107 and 108. Notice the marked difference between the two figures, both of which show the same set of pulleys, and the last figure, by merely having the lower and the right-hand lines of each pulley heavy, changes the character of the representation, and tells much more clearly what the draughtsman sought to convey.

Scale Drawings.—All drawings are made to a p. 101 scale where the article is large and cannot be indicated the exact size, using parts of an inch to represent inches; and parts of a foot to represent feet.

In order to reduce a drawing where a foot is the unit, it is always best to use one-and-a-half inches, or twelve-eighths of an inch, as the basis. In this way each eighth of an inch represents an inch. If the drawing should be made larger, then use three inches, and in that way each inch would be one-quarter of an inch.

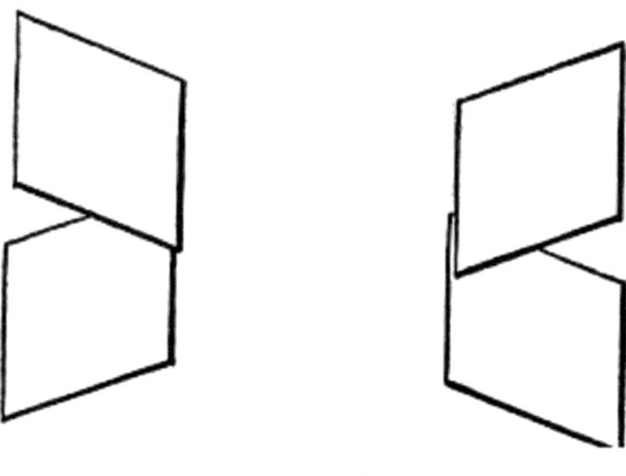

Fig. 109.   Fig. 110.
Illustrating Heavy Lines ToList

The drawing should then have marked, in some conspicuous place, the scale, like the following: "Scale, 1 1/2" = 1'"; or, "Scale 3" = 1'."

Degree, and What it Means.—A degree is not p. 102 a measurement. The word is used to designate an interval, a position, or an angle. Every circle has 360 degrees, and when a certain degree is mentioned, it means a certain angle from what is called a *base line*.

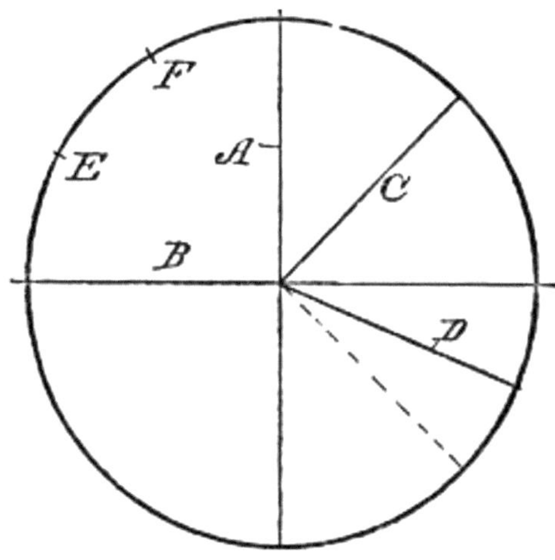

Fig.

111. Illustrating Degrees ToList

Look at Fig. 111. This has a vertical line A, and a horizontal line B. The circle is thus divided into four parts, and where these lines A, B, cross the circle are the cardinal points. Each of the four parts is called a quadrant, and each quadrant has 90 degrees.

Any line, like C, which is halfway between A and B, is 45 degrees. Halfway between A and C, p. 103 or between B and C, like the line D, is 22 1/2 degrees.

Memorizing Angles.—It is well to try and remember these lines by fixing the angles in the memory. A good plan is to divide any of the quadrants into thirds, as shown by the points E, F, and then remember that E is 30 degrees from the horizontal line B, and that F is 60 degrees. Or, you might say that F is 30 degrees from the vertical line A, and E 60 degrees from A. Either would be correct.

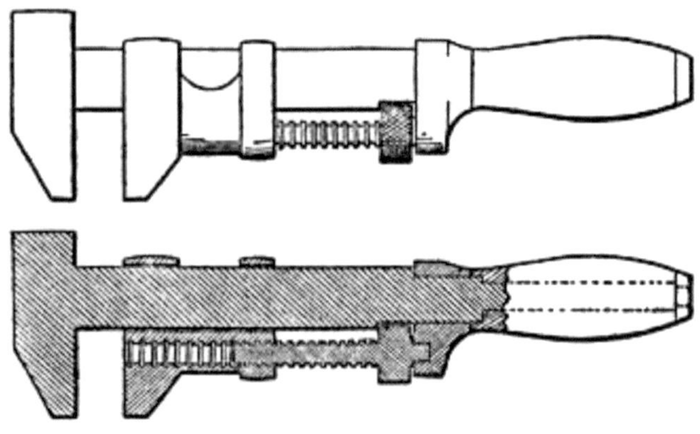

112. Section Lining ToList

Section Lining. — In representing many parts of a machine, or article, it is necessary to show the parts cut off, which must be illustrated by what is called "section lining." Adjacent parts should p. 104 have the section lines running at right angles to each other, and always at 45 degrees.

Look at the outside and then the inside views of Fig. 112, and you will see how the contiguous parts have the angles at right angles, and clearly illustrate how every part of the wrench is made. Skill in depicting an article, for the purpose of constructing it from the drawing, will make the actual work on the bench and lathe an easy one.

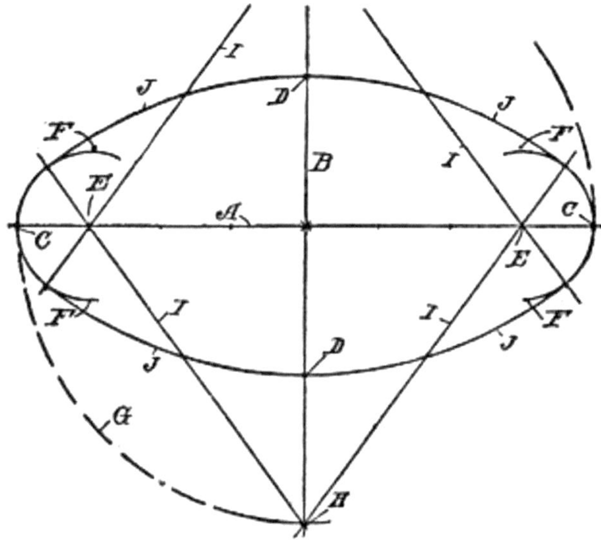

113. Drawing an Ellipse ToList

Making Ellipses and Irregular Curves. — This is the hardest thing to do with drawing tools. A properly constructed elliptical figure is difficult, p. 105 principally, because two different sized curves are required, and the pen runs from one curve into the other. If the two curves meet at the wrong place, you may be sure you will have a distorted ellipse.

Follow the directions given in connection with Fig. 113, and it will give you a good idea of merging the two lines.

First. Draw a horizontal line, A, which is in the direction of the major axis of the ellipse — that is, the longest distance across. The narrow part of the ellipse is called the minor axis.

Second. Draw a perpendicular line, B, which we will call the center of the ellipse, where it crosses the line A. This point must not be confounded with the *focus*. In a circle the focus is the exact center of the ring, but there is no such thing in an ellipse. Instead, there are two focal points, called the *foci*, as you will see presently.

Third. Step off two points or marking places, as we shall term them, equidistant from the line B, and marked C, C. These marks will then represent the diameter of the ellipse across its major axis.

Fourth. We must now get the diameter of the minor axis, along the line B. This distance will depend on the perspective you have of the figure. If you look at a disk at an angle of about 30 degrees it will be half of the distance across the major axis

p. 106

So you may understand this examine Fig. 114. The first sketch shows the eye looking directly at the disk 1. In the second sketch the disk is at 30 degrees, and now the lines 2 2, from the eye, indicate that it is just half the width that it was when the lines 3 3 were projected. The marks D D, therefore, indicate the distance across the minor axis in Fig. 113.

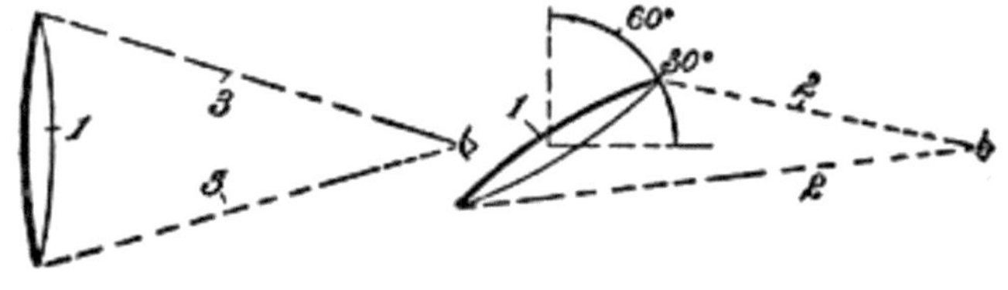

Fig.

114. Perspection in Angles ToList

Fifth. We must now find the focal points of the ellipse. If the line A on each side of the cross line B is divided into four parts, the outer marks E may be used for the foci, and will be the places where the point of the compass, or bow pen, is to be placed.

Sixth. Describe a circle F, so it passes through the mark C, and move the point of the compass to the center of the ellipse, at the star, and describe a circle line G, from the mark C to the line B. This will give a centering point H. Then draw a line I from H to E, and extend it through the circle F.

Seventh. If the point of the compass is now put p. 107 at H, and the pencil or pen on the circle line F, the curve J can be drawn, so the latter curve and the curve F will thus merge perfectly at the line I.

The Focal Points. — The focal points can be selected at any arbitrary point, between C and the line B, and the point H may be moved closer to or farther away from the line A, and you will succeed in making the ellipse correct, if you observe one thing, namely: The line I, which must always run from H to E, and intersects the circle F, is the starting or the ending point for the small curve F or the large circle J.

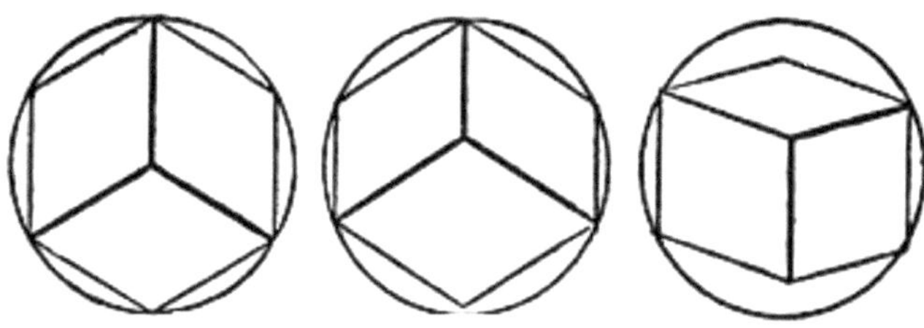

Figs. 115-117. Perspectives of Cubes ToList

Isometric and Perspective. — A figure may be drawn so as to show an isometric or a perspective view. Thus, a cube can be drawn so as to make an isometric figure, as in Fig. 115, where the three sides are equal to each other.

Isometric means a method of drawing any object in such a manner that the height, length and p. 108 breadth may be shown in the proportion they really bear to each other. Fig. 115 has the sides not only equal to each other, in appearance to the eye, but they have the same outlines and angles.

Contrast this figure with Figs. 116 and 117. In Fig. 116 two of the sides are equal in angles and outline; and in Fig. 117 each side has a different outline, and different angles. Nevertheless, all the cubes are, in reality, of the same dimension.

The Protractor. — This is a most useful tool for the draughtsman. It enables the user to readily find any angle. Fig. 118 shows an approved form of the tool for this purpose.

Fig.

118. Protractor.
Section Lining Metals ToList

Suggestions in Drawing.—As in the use of all p. 109 other tools, so with the drawing instrument, it must be kept in proper order. If the points are too fine they will cut the paper; if too blunt the lines will be ragged. In whetting the points hold the pen at an angle of 12 degrees. Don't make too long an angle or slope, and every time you sharpen hold it at the same angle, so that it is ground back, and not at the point only.

119. Using the Protractor. ToList

Holding the Pen. — The drawing pen should be held as nearly vertical as possible. Use the cleaning rag frequently. If the ink does not flow freely, after you have made a few strokes, as is frequently the case, gently press together the points. The least grit between the tines will cause an irregular flow

p. 110

Inks. — As prepared liquid inks are now universally used, a few suggestions might be well concerning them. After half the bottle has been used, add a half teaspoonful of water, shake it well, and then strain it through a fine cotton cloth. This will remove all grit and lint that is sure to get into the bottle however carefully it may be corked.

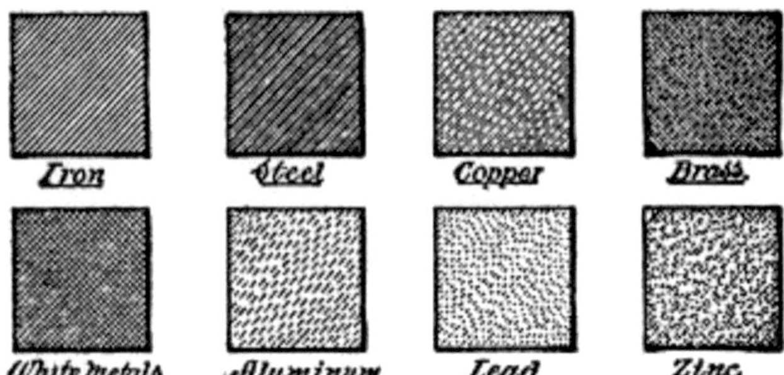

Fig. 120. Section Lining Metals ToList

Tracing Cloth.—It is preferable to use the dull side of the tracing cloth for the reasons that, as the cloth is rolled with the glossy side inside, the figure when drawn on the other side will be uppermost, and will thus lie flat; and on the other hand, the ink will take better on the dull side.

If the ink does not flow freely, use chalk, fine pumice stone, or talc, and rub it in well with a clean cloth, and then wipe off well before beginning to trace

p. 111

Detail Paper.—The detail paper, on which the drawing is first made in pencil, should show the figure accurately, particularly the points where the bow pen are to be used, as well as the measurement points for the straight lines.

How to Proceed.—Make the circles, curves, and irregular lines first, and then follow with the straight lines. Where the point of the circle pen must be used for a large number of lines, as, for instance, in shading, the smallest circles should be made first, and the largest circles last, because at every turn the centering hole becomes larger, and there is liability to make the circles more or less irregular. Such irregularity will not be so noticeable in the large curves as in the smaller ones.

Indicating Material by the Section Lines.—In section lining different materials can be indicated by the character of the lines, shown in Fig. 120.

p. 112

TREATMENT AND USE OF METALS

Annealing.—A very important part of the novice's education is a knowledge pertaining to the annealing of metals. Unlike the artisan in wood, who works the materials as he finds them, the machinist can, and, in fact, with many of the substances, must prepare them so they can be handled or cut by the tools.

Annealing is one of the steps necessary with all cutting tools, and it is an absolute requirement with many metals for ordinary use, as well as for many other articles like glass. This is particularly true in the use of copper.

Toughness and Elasticity.—It means the putting of metals in such a condition that they will not only be less brittle, but also tougher and more elastic. Many substances, like glass, must be annealed before they can be put in condition for use, as this material when first turned out is so brittle that the slightest touch will shatter it, so that it must be toughened.

Malleable or wrought iron, if subjected to pressure, becomes brittle, and it is necessary to anneal it. Otherwise, if used, for instance, for boiler p. 113 plates, from the rolled sheets, it would stand but little pressure.

The most immediate use the boy will have is the treatment of steel. He must learn the necessity of this process, and that of tempering, in all his cutting tools, and in the making of machinery where some parts are required to be constructed of very hard metal.

The Process.—To anneal steel it must be heated to a bright cherry red and then gradually cooled down. For this purpose a bed of fine charcoal, or iron filings and lime, is prepared, in which the article is embedded, and permitted to remain until it is cold.

There are many ways of doing the work, particularly in the use of substances which will the most readily give up their carbon to the tool. Yellow prussiate of potash is an excellent medium, and this is sprinkled over the cherry-heated article to be annealed. The process may be repeated several times.

Tempering.—This is the reverse of annealing as understood in the art. The word itself does not mean to "harden," but to put into some intermediate state. For instance, "tempered clay" means a clay which has been softened so it can be readily worked.

On the other hand, a tempered steel tool is put into a condition where it is hardened, but this hard p. 114 ness is also accompanied by another quality, namely, *toughness*. For this reason, the word *temper*, and not *hardness*, is referred to. A lathe tool, if merely hardened, would be useless for that purpose.

Tempering Contrasted with Annealing.—It will be observed that in annealing three things are necessary: First, heating to a certain temperature; second, cooling slowly; third, the particular manner of cooling it.

In tempering, on the other hand, three things are also necessary:

First: The heating temperature should be a dull red, which is less than the annealing heat.

Second: Instead of cooling slowly the article tempered is dipped into a liquid which suddenly chills it.

Third: The materials used vary, but if the article is plunged into an unguent made of mercury and bacon fat, it will impart a high degree of toughness and elasticity.

Materials Used.—Various oils, fats and rosins are also used, and some acids in water are also valuable for this purpose. Care should be taken to have sufficient amount of liquid in the bath so as not to evaporate it or heat it up too much when it receives the heated body.

Different parts of certain articles require varying degrees of hardness, like the tangs of files. The p. 115 cutting body of the file must be extremely hard, and rather brittle than tough. If the tang should be of the same hardness it would readily break.

Gradual Tempering.—To prevent this, some substance like soap suds may be used to cool down the tang, so that toughness without hardness is imparted.

The tempering, or hardening, like the annealing process, may be repeated several times in succession, and at each successive heating the article is put at a higher temperature.

If any part of a body, as, for instance, a hammerhead, should require hardening, it may be plunged into the liquid for a short distance only, and this will harden the pole or peon while leaving the other part of the head soft, or annealed.

Glycerine is a good tempering substance, and to this may be added a small amount of sulphate of potash.

Fluxing.—The word *flux* means to fuse or to melt, or to put into a liquid state. The office of a flux is to facilitate the fusion of metals. But fluxes do two things. They not only aid the conversion of the metal into a fluid state, but also serve as a means for facilitating the unity of several metals which make up the alloy, and aid in uniting the parts of metals to be joined in the welding of parts

p. 116

Uniting Metals.—Metals are united in three ways, where heat is used:

First: By heating two or more of them to such a high temperature that they melt and form a compound, or an alloy, as it is called.

Second: By heating up the points to be joined, and then lapping the pieces and hammering the parts. This is called forge work or welding.

Third: By not heating the adjacent parts and using an easily fusible metal, which is heated up and run between the two, by means of a soldering iron.

The foreign material used in the first is called a flux; in the second it is termed a welding compound; and in the third it is known as a soldering acid, or soldering fluid.

The boy is not so much interested in the first process, from the standpoint of actual work, but it is necessary that he should have some understanding of it.

It may be said, as to fluxes, generally, that they are intended to promote the fusion of the liquefying metals, and the elements used are the alkalis, such as borax, tartar, limestone, or fluor spar.

These substances act as reducing or oxidizing agents. The most important are carbonate of soda, potash, and cyanide of potassium. Limestone is used as the flux in iron-smelting

p. 117

Welding Compounds.—Elsewhere formulas are given of the compounds most desirable to use. It is obvious that the application of these substances on the heated surfaces, is not only to facilitate the heating, but to prepare the articles in such a manner that they will more readily adhere to each other.

Oxidation.—Oxidation is the thing to guard against in welding. The moment a piece of metal, heated to whiteness, is exposed, the air coats it with a film which is called an *oxide*. To remove this the welding compound is applied.

The next office of the substance thus applied, is to serve as a medium for keeping the welding parts in a liquid condition as long as possible, and thus facilitate the unity of the joined elements.

When the hammer beats the heated metals an additional increment of heat is imparted to the weld, due to the forcing together of the molecules of the iron, so that these two agencies, namely, the compound and the mechanical friction, act together to unite the particles of the metal.

Soldering.—Here another principle is involved, namely, the use of an intermediate material between two parts which are to be united. The surfaces to be brought together must be thoroughly cleaned, using such agents as will prevent the formation of oxides.

The parts to be united may be of the same, or of p. 118 different materials, and it is in this particular that the workman must be able to make a choice of the solder most available, and whether hard or soft.

Soft Solder.—A soft solder is usually employed where lead, tin, or alloys of lead, tin and bismuth are to be soldered. These solders are all fusible at a low temperature, and they do not, as a result, have great strength.

Bismuth is a metal which lowers the fusing point of any alloy of which it forms a part, while lead makes the solder less fusible.

Hard Solder.—These are so distinguished because they require a temperature above the low red to fuse them. The metals which are alloyed for this purpose are copper, silver, brass, zinc and tin. Various alloys are thus made which require a high temperature to flux properly, and these are the ones to use in joining steel to steel, the parts to be united requiring an intense furnace heat.

Spelter.—The alloy used for this purpose is termed "spelter," and brass, zinc and tin are its usual components. The hard solders are used for uniting brass, bronze, copper, and iron.

Whether soft or hard solder is used, it is obvious that it must melt at a lower temperature than the parts which are to be joined together.

There is one peculiarity with respect to alloys: p. 119 They melt at a lower temperature than either of the metals forming the alloys.

Soldering Acid.—Before beginning the work of soldering, the parts must be cleaned by filing or sandpapering, and coated with an acid which neutralizes the oxygen of the air.

This is usually muriatic acid, of which use, say, one quart and into this drop small pieces of zinc. This will effervesce during the time the acid is dissolving the zinc. When the boiling motion ceases, the liquid may be strained, or the dark pieces removed.

The next step is to dissolve two ounces of sal ammoniac in a third of a pint of water, and in another vessel dissolve an ounce of chloride of tin.

Then mix the three solutions, and this can be placed in a bottle, or earthen jar or vessel, and it will keep indefinitely.

The Soldering Iron.—A large iron is always better than a small one, particularly for the reason that it will retain its heat better. This should always be kept tinned, which can be done by heating and plunging it into the soldering solution, and the solder will then adhere to the iron and cover the point, so that when the actual soldering takes place the solder will not creep away from the tool.

By a little care and attention to these details, the work of uniting metals will be a pleasure. It is so p. 120 often the case, however, that the apparatus for doing this work is neglected in a shop; the acid is allowed to become dirty and full or foreign matter, and the different parts separated

p. 121

CHAPTER X ToC

ON GEARING AND HOW ORDERED

The technical name for gears, the manner of measuring them, their pitch and like terms, are most confusing to the novice. As an aid to the understanding on this subject, the wheels are illustrated, showing the application of these terms.

Spur and Pinion.—When a gear is ordered a specification is necessary. The manufacturer will know what you mean if you use the proper terms, and you should learn the distinctions between spur and pinion, and why a bevel differs from a miter gear.

If the gears on two parallel shafts mesh with each other, they both may be of the same diameter, or one may be larger than the other. In the latter case, the small one is the pinion, and the larger one the spur wheel.

Some manufacturers use the word "gear" for "pinion," so that, in ordering, they call them *gear* and *pinion*, in speaking of the large and small wheels.

Measuring a Gear.—The first thing to specify would be the diameter. Now a spur gear, as well as a pinion, has three diameters; one measure p. 122 across the outer extremities of the teeth; one measure across the wheel from the base of the teeth; and the distance across the wheel at a point midway between the base and end of the teeth.

These three measurements are called, respectively, "outside diameter," "inside diameter," and "pitch diameter." When the word *diameter* is used, as applied to a gear wheel, it is always understood to mean the "pitch diameter."

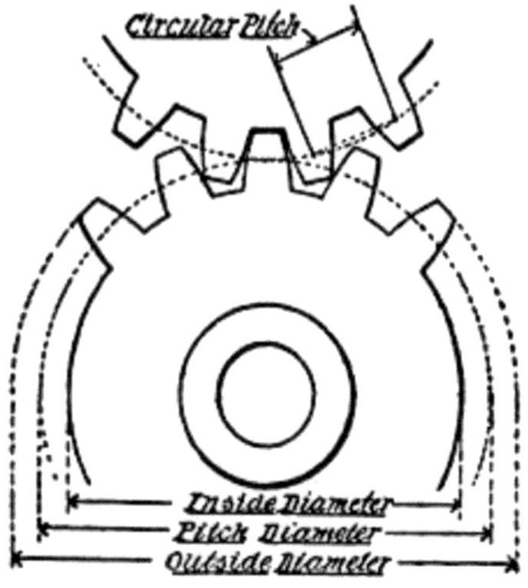

Fig.

121. Spur Gears ToList

Pitch. — This term is the most difficult to understand. When two gears of equal size mesh together, the pitch line, or the *pitch circle*, as it is p. 123 also called, is exactly midway between the centers of the two wheels.

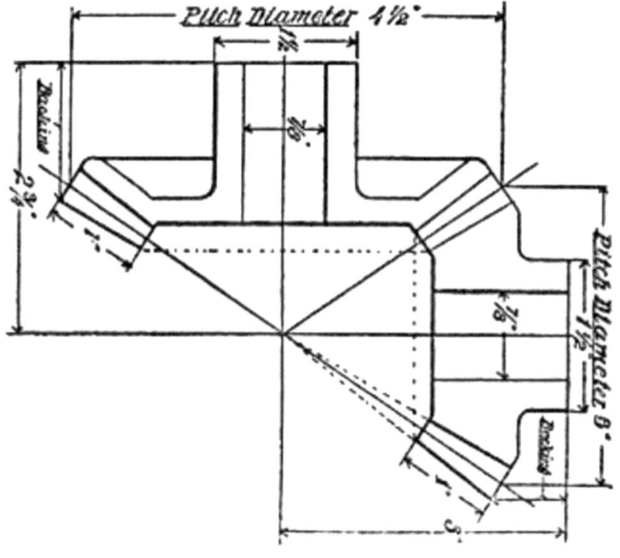

Fig.

122. Miter Gear Pitch ToList

Now the number of teeth in a gear is calculated on the pitch line, and this is called:

Diametral Pitch.—To illustrate: If a gear has 40 teeth, and the pitch diameter of the wheel is 4 inches, there are 10 teeth to each inch of the pitch diameter, and the gear is then 10 *diametral pitch.*

Circular Pitch.—Now the term "circular pitch" grows out of the necessity of getting the measurement of the distance from the center of one tooth p. 124 to the center of the next, and it is measured along the pitch line.

Supposing you wanted to know the number of teeth in a gear where the pitch diameter and the diametral pitch are given. You would proceed as follows: Let the diameter of the pitch circle be 10 inches, and the diameter of the diametral pitch be 4 inches. Multiplying these together the product is 40, thus giving the number of teeth.

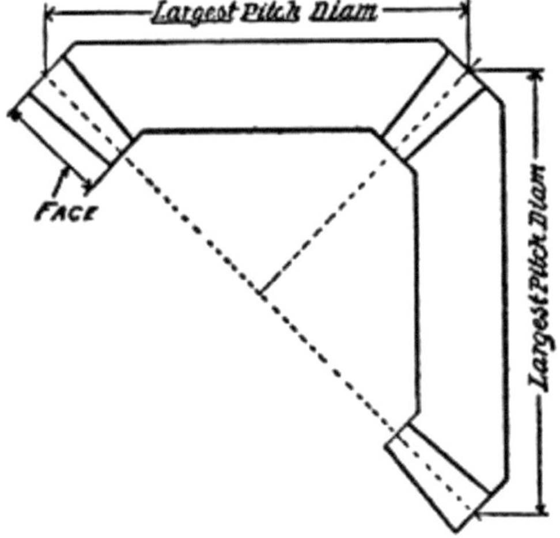

Fig.

123. Bevel Gears. ToList

It will thus be seen that if you have an idea of the diametral pitch and circular pitch, you can pretty fairly judge of the size that the teeth will be, and thus enable you to determine about what kind of teeth you should order

p. 125

How to Order a Gear.—In proceeding to order, therefore, you may give the pitch, or the diameter of the pitch circle, in which latter case the manufacturer of the gear will understand how to determine the number of the teeth. In case the intermeshing gears are of different diameters, state the number of teeth in the gear and also in the pinion, or indicate what the relative speed shall be.

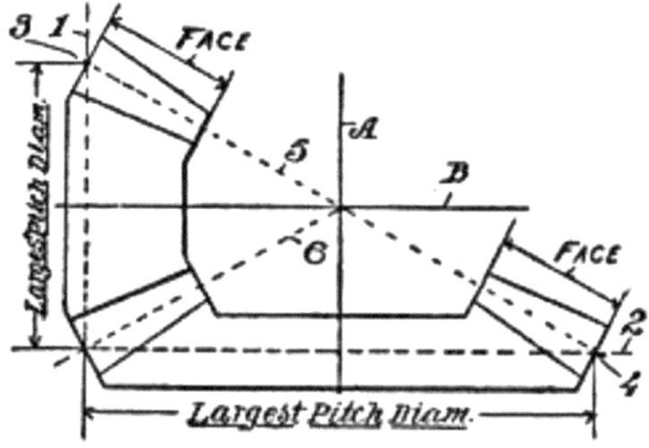

Fig.

124. Miter Gears. ToList

This should be followed by the diameter of the hole in the gear and also in the pinion; the backing of both gear and pinion; the width of the face; the diameter of the gear hub; diameter of the pinion hub; and, finally, whether the gears are to be fastened to the shafts by key-ways or set-screws.

Fig. 122 shows a sample pair of miter gears, p. 126 with the measurements to indicate how to make the drawings. Fig. 123 shows the bevel gears.

Bevel and Miter Gears.—When two intermeshing gears are on shafts which are at right angles to each other, they may be equal diametrically, or of different sizes. If both are of the same diameter, they are called bevel gears; if of different diameters, miter gears.

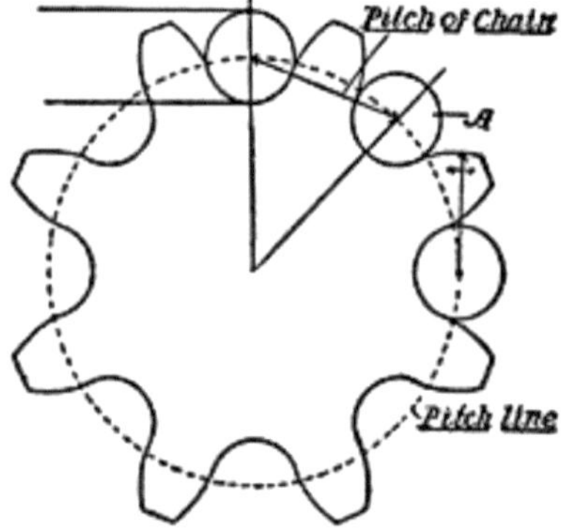

Fig.

125. Sprocket Wheel. ToList

It is, in ordering gears of this character, that the novice finds it most difficult to know just what to do. In this case it is necessary to get the proper relation of speed between the two gears, and, for convenience, we shall, in the drawing, make the gears in the relation of 2 to 1.

Drawing Gears.—Draw two lines at right angles, p. 127 Fig. 124, as 1 and 2, marking off the sizes of the two wheels at the points 3, 4. Then draw a vertical line (A) midway between the marks of the line 2, and this will be the center of the main pinion.

Also draw a horizontal line (B) midway between the marks on the vertical line (1), and this will represent the center of the small gear. These two cross lines (A, B) constitute the intersecting axes of the two wheels, and a line (5), drawn from the mark (3 to 4), and another line (6), from the axes to the intersecting points of the lines (1, 2), will give the pitch line angles of the two wheels.

Sprocket Wheels.—For sprocket wheels the pitch line passes centrally through the rollers (A) of the chain, as shown in Fig. 125, and the pitch of the chain is that distance between the centers of two adjacent rollers. In this case the cut of the teeth is determined by the chain

p. 128

CHAPTER XI ToC

MECHANICAL POWERS

The Lever.—The lever is the most wonderful mechanical element in the world. The expression, *lever*, is not employed in the sense of a stick or a bar which is used against a fulcrum to lift or push something with, but as the type of numerous devices which employ the same principle.

Some of these devices are, the wedge, the screw, the pulley and the inclined plane. In some form or other, one or more of these are used in every piece of mechanism in the world.

Because the lever enables the user to raise or move an object hundreds of times heavier than is possible without it, has led thousands of people to misunderstand its meaning, because it has the appearance, to the ignorant, of being able to manufacture power.

Wrong Inferences from Use of Lever.—This lack of knowledge of first principles, has bred and is now breeding, so-called perpetual motion inventors (?) all over the civilized world. It is surprising how many men, to say nothing of boys, actually believe that power can be made without the expenditure of something which equalizes it

p. 129

The boy should not be led astray in this particular, and I shall try to make the matter plain by using the simple lever to illustrate the fact that whenever power is exerted some form of energy is expended.

In Fig. 126 is a lever (A), resting on a fulcrum (B), the fulcrum being so placed that the lever is four times longer on one side than on the other. A weight (C) of 4 pounds is placed on the short end, and a 1-pound weight (D), called the *power*, on the short end. It will thus be seen that the lever is balanced by the two weights, or that the *weight* and the *power* are equal.

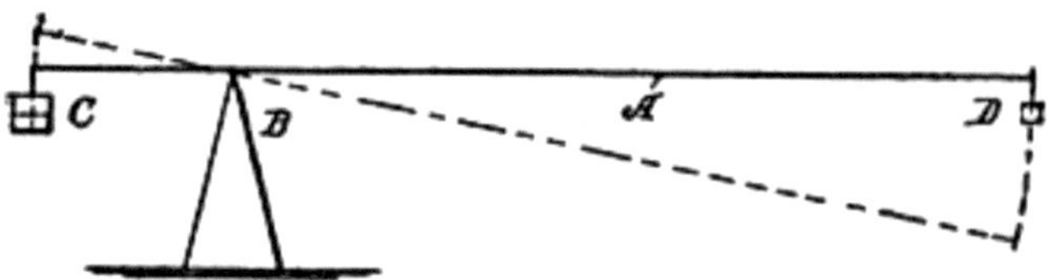

Fig. 126. Simple Lever ToList

The Lever Principle.—Now, without stopping to inquire, the boy will say: "Certainly, I can understand that. As the lever is four times longer on one side of the fulcrum than on the other side, it requires only one-fourth of the weight to balance the four pounds. But suppose I push down the lever, at the point where the weight (D) is, p. 130 then, for every pound I push down I can raise four pounds at C. In that case do I not produce four times the power?"

I answer, yes. But while I produce that power I am losing something which is equal to the power gained. What is that?

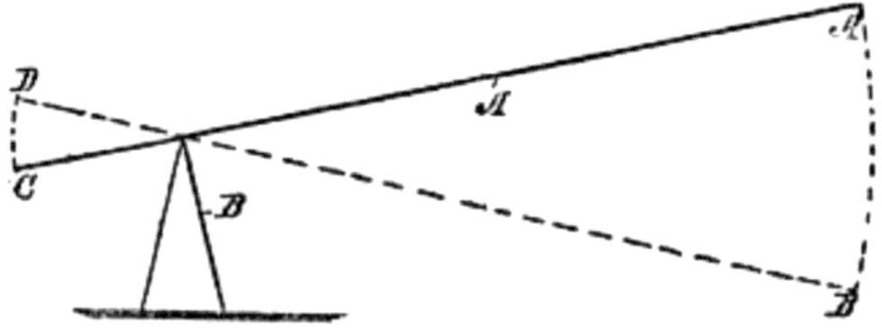

Fig. 127. Lever Action ToList

First: Look at Fig. 127; the distance traveled. The long end of the lever is at its highest point, which is A; and the short end of the lever is at its lowest point C. When the long end of the lever is pushed down, so it is at B, it moves four times farther than the short end moves upwardly, as the distance from C to D is just one-fourth that from A to B. The energy expended in moving four times the distance balances the power gained.

Power vs. Distance Traveled.—From this the following law is deduced: That whatever is gained in power is lost in the distance traveled

Second: Using the same figure, supposing it was necessary to raise the short end of the lever, from C to D, in one second of time. In that case the hand pressing down the long end of the lever, would go from A to B in one second of time; or it would go four times as far as the short end, in the same time.

Power vs. Loss in Time.—This means another law: That what is gained in power is lost in time.

Distinguish clearly between these two motions. In the first case the long end of the lever is moved down from A to B in four seconds, and it had to travel four times the distance that the short end moves in going from C to D.

In the second case the long end is moved down, from A to B, in one second of time, and it had to go that distance in one-fourth of the time, so that four times as much energy was expended in the same time to raise the short end from C to D.

Wrongly Directed Energy.—More men have gone astray on the simple question of the power of the lever than on any other subject in mechanics. The writer has known instances where men knew the principles involved in the lever, who would still insist on trying to work out mechanical devices in which pulleys and gearing were involved, without seeming to understand that those p. 132 mechanical devices are absolutely the same in principle.

This will be made plain by a few illustrations. In Fig. 128, A is a pulley four times larger, diametrically, than B, and C is the pivot on which they turn. The pulleys are, of course, secured to each other. In this case we have the two weights, one of four pounds on the belt, which is on the small pulley (B), and a one-pound weight on the belt from the large pulley (A).

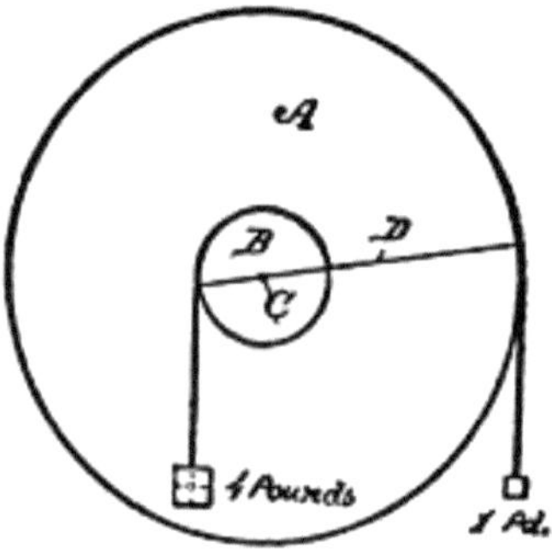

Fig.

128. The Pulley ToList

The Lever and the Pulley.—If we should substitute a lever (D) for the pulleys, the similarity to the lever (Fig. 127) would be apparent at once. The pivot (C) in this case would act the same as the pivot (C) in the lever illustration.

In the same manner, and for like reasons, the p. 133 wedge, the screw and the incline plane, are different structural applications of the principles set forth in the lever.

Whenever two gears are connected together, the lever principle is used, whether they are the same in size, diametrically, or not. If they are the same size then no change in power results; but instead, thereof, a change takes place in the direction of the motion.

Fig. 129.   Fig. 130.

Change of Direction ToList

When one end of the lever (A) goes down, the other end goes up, as shown in Fig. 129; and in Fig. 130, when the shaft (C) of one wheel turns in one direction, the shaft of the other wheel turns in the opposite direction.

140

It is plain that a gear, like a lever, may change direction as well as increase or decrease power. It is the thorough knowledge of these facts, and their application, which enables man to make the wonderful machinery we see on every hand.

Sources of Power.—Power is derived from a p. 134 variety of sources, but what are called the *prime movers* are derived from heat, through the various fuels, from water, from the winds and from the tides and waves of the ocean. In the case of water the power depends on the head, or height, of the surface of the water above the discharging orifice.

Water Power.—A column of water an inch square and 28 inches high gives a pressure at the base of one pound; and the pressure at the lower end is equal in all directions. If a tank of water 28 inches high has a single orifice in its bottom 1" x 1" in size, the pressure of water through that opening will be only one pound, and it will be one pound through every other orifice in the bottom of the same size.

Calculating Fuel Energy.—Power from fuels depends upon the expansion of the materials consumed, or upon the fact that heat expands some element, like water, which in turn produces the power. One cubic inch of water, when converted into steam, has a volume equal to one cubic foot, or about 1,700 times increase in bulk.

Advantage is taken of this in steam engine construction. If a cylinder has a piston in it with an area of 100 square inches, and a pipe one inch square supplies steam at 50 pounds pressure, the piston will have 50 pounds pressure on every square inch of its surface, equal to 5,000 pounds

p. 135

The Pressure or Head.—In addition to that there will also be 50 pounds pressure on each square inch of the head, as well as on the sides of the cylinder.

Fig. 131 shows a cylinder (A), a piston (B) and a steam inlet port (C), in which is indicated how the steam pressure acts equally in all directions. As, however, the piston is the only movable part, the force of the steam is directed to that part, and the motion is then transmitted to the crank, and to the shaft of the engine.

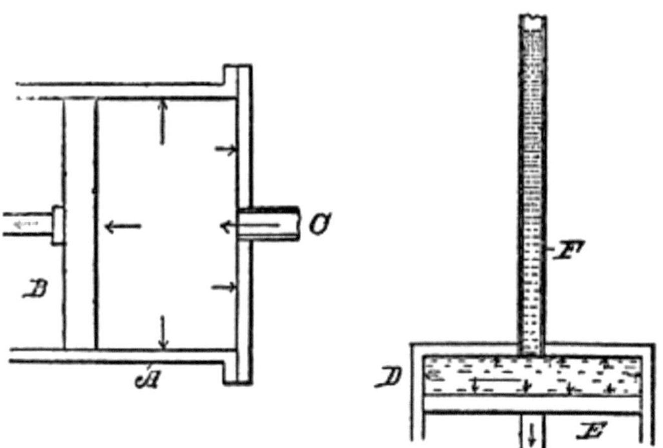

Fig. 131. Steam Pressure   Fig. 132. Water Pressure ToList

This same thing applies to water which, as stated, is dependent on its head. Fig. 132 repre p. 136 sents a cylinder (D) with a vertically movable piston (E) and a standpipe (F). Assuming that the pipe (F) is of sufficient height to give a pressure of 50 pounds to the square inch, then the piston (E) and the sides and head of the cylinder (D) would have 50 pounds pressure on every square inch of surface.

Fuels. — In the use of fuels, such as the volatile hydrocarbons, the direct expansive power of the fuel gases developed, is used to move the piston back and forth. Engines so driven are called *Internal Combustion Motors*.

Power from Winds. — Another source of power is from the wind acting against wheels which have blades or vanes disposed at such angles that there is a direct conversion of a rectilinear force into circular motion.

In this case power is derived from the force of the moving air and the calculation of energy developed is made by considering the pressure on each square foot of surface. The following table shows the force exerted at different speeds against a flat surface one foot square, held so that the wind strikes it squarely: p. 137

SPEED OF WIND			PRESSURE			SPEED OF WIND			PRESSURE		
5	Miles per hour			2	oz.	35	miles per hour		6	lb.	2 oz.
10	"		88	"		40	"		"	8	"
15	"	1 lb.	2	"		45	"		"	10	"	2 "
20	"		2	"	"	50	"		"	12	"	2 "
25	"		3	"	2 "	55	"		"	15	"	2 "
30	"		4	"	8 "	60	"		"	18	"	

Varying Degrees of Pressure.—It is curious to notice how the increase in speed changes the pressure against the blade. Thus, a wind blowing 20 miles an hour shows 2 pounds pressure; whereas a wind twice that velocity, or 40 miles an hour, shows a pressure of 8 pounds, which is four times greater than at 20 miles.

It differs, therefore, from the law with respect to water pressure, which is constant in relation to the height or the head—that is, for every 28 inches height of water a pound pressure is added.

Power from Waves and Tides.—Many attempts have been made to harness the waves and the tide and some of them have been successful. This effort has been directed to the work of converting the oscillations of the waves into a rotary motion, and also to take advantage of the to-and-fro movement of the tidal flow. There is a great field in this direction for the ingenious boy.

A Profitable Field.—In no direction of human p. 138 enterprise is there such a wide and profitable field for work, as in the generation of power. It is constantly growing in prominence, and calls for the exercise of the skill of the engineer and the ingenuity of the mechanic. Efficiency and economy are the two great watchwords, and this is what the world is striving for. Success will come to him who can contribute to it in the smallest degree.

Capital is not looking for men who can cheapen the production of an article 50 per cent., but 1 per cent. The commercial world does not expect an article to be 100 per cent, better. Five per cent. would be an inducement for business

p. 139

CHAPTER XII ToC

ON MEASURES

Horse-power.—When work is performed it is designated as horse-power, usually indicated by the letters H. P.; but the unit of work is called a *foot pound*.

If one pound should be lifted 550 feet in one second, or 550 pounds one foot in the same time, it would be designated as one horse-power. For that reason it is called a foot pound. Instead of using the figure to indicate the power exerted during one minute of time, the time is taken for a minute, in all calculations, so that 550 multiplied by the number of seconds, 60, in a minute, equals 33,000 foot pounds.

Foot Pounds.—The calculation of horse-power is in a large measure arbitrary. It was determined in this way: Experiments show that the heat expended in vaporizing 34 pounds of water per hour, develops a force equal to 33,000 foot pounds; and since it takes about 4 pounds of coal per hour to vaporize that amount of water, the heat developed by that quantity of coal develops the same force as that exercised by an average horse exerting his strength at ordinary work

p. 140

All power is expressed in foot pounds. Suppose a cannon ball of sufficient weight and speed strikes an object. If the impact should indicate 33,000 pounds it would not mean that the force employed was one horse-power, but that many foot pounds.

If there should be 60 impacts of 550 pounds each within a minute, it might be said that it would be equal to 1 horse-power, but the correct way to express it would be foot pounds.

So in every calculation, where power is to be calculated, first find out how many foot pounds are developed, and then use the unit of measure, 33,000, as the divisor to get the horse-power, if you wish to express it in that way.

It must be understood, therefore, that horse-power is a simple unit of work, whereas a foot pound is a compound unit formed of a foot paired with the weight of a pound.

Energy.—Now *work* and *energy* are two different things. Work is the overcoming of resistance of any kind, either by causing or changing motion, or maintaining it against the action of some other force.

Energy, on the other hand, is the power of doing work. Falling water possesses energy; so does a stone poised on the edge of a cliff. In the case of water, it is called *kinetic* energy; in the stone p. 141 *potential* energy. A pound of pressure against the stone will cause the latter, in falling, to develop an enormous energy; so it will be seen that this property resides, or is within the thing itself. It will be well to remember these definitions.

How to Find Out the Power Developed.—The measure of power produced by an engine, or other source, is so interesting to boys that a sketch is given of a Prony Brake, which is the simplest form of the Dynamometer, as these measuring machines are called.

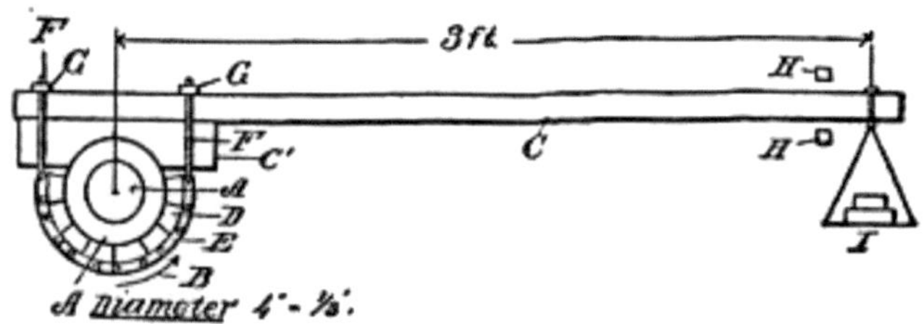

Fig.

133. Prony Brake ToList

In the drawing (A) is the shaft, with a pulley (A´), which turns in the direction of the arrow (B). C is a lever which may be of any length. This has a block (C´), which fits on the pulley, and below the shaft, and surrounding it, are blocks (D) held against the pulley by a chain (E), the ends of the chain being attached to bolts (F) which pass through the block (C´) and lever (C)

p. 142

Nuts (G) serve to draw the bolts upwardly and thus tighten the blocks against the shaft. The free end of the lever has stops (H) above and below, so as to limit its movement. Weights (I) are suspended from the end of the lever.

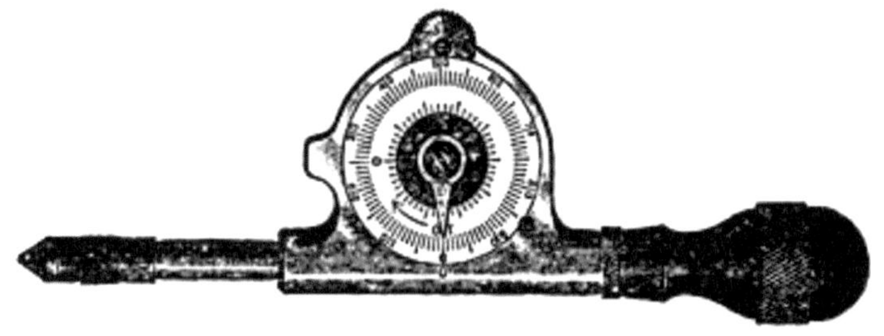

Fig. 134. Speed Indicator ToList

The Test. — The test is made as follows: The shaft is set in motion, and the nuts are tightened until its full power at the required speed is balanced by the weight put on the platform.

The following calculation can then be made:

For our present purpose we shall assume that the diameter of the pulley (A´) is 4 inches; the length of the lever (C), 3 feet; the speed of the shaft (A) and the pulley, 210 revolutions per minute; and the weight 600 pounds.

Now proceed as follows:

(1) Multiply the diameter of the pulley (A´) p. 143 (4 inches) by 3.1416, and this will give the circumference 12.5664 inches; or, 1.0472 feet.

(2) Multiply this product (1.0472) by the revolutions per minute. $1.0472 \times 210 = 219.912$. This equals the *speed* of the periphery of the pulley.

(3) The next step is to get the length of the lever (C) from the center of the shaft (A) to the point from which the weights are suspended, and divide this by one-half of the diameter of the pulley (A´). $36" \div 2" = 18"$, or 1 1/2 feet. This is the *leverage*.

(4) Then multiply the *weight* in pounds by the *leverage*. 600 × 1 1/2 = 900.

(5) Next multiply this product (900) by the *speed*, 900 × 219.912 = 197,920.8, which means *foot pounds*.

(6) As each horse-power has 33,000 foot pounds, the last product should be divided by this figure, and we have 197,920.8 ÷ 33,000 = 5.99 H. P.

The Foot Measure.—How long is a foot, and what is it determined by? It is an arbitrary measure. The human foot is the basis of the measurement. But what is the length of a man's foot? It varied in different countries from 9 to 21 inches.

In England, in early days, it was defined as a measure of length consisting of 12 inches, or 36 barleycorns laid end to end. But barleycorns differ in length as well as the human foot, so the p. 144 standard adopted is without any real foundation or reason.

Weight.—To determine weight, however, a scientific standard was adopted. A gallon contains 8.33 pounds avoirdupois weight of distilled water. This gallon is divided up in two ways; one by weight, and the other by measurement.

Each gallon contains 231 cubic inches of distilled water. As it has four quarts, each quart has 57 3/4 cubic inches, and as each quart is comprised of two pints, each pint has nearly 29 cubic inches.

The Gallon.—The legal gallon in the United States is equal to a cylindrical measure 7 inches in diameter and 6 inches deep.

Notwithstanding the weights and dimensions of solids and liquids are thus fixed by following a scientific standard, the divisions into scruples, grains, pennyweights and tons, as well as cutting them up into pints, quarts and other units, is done without any system, and for this reason the need of a uniform method has been long considered by every country.

The Metric System.—As early as 1528, Fernal, a French physician, suggested the metric system. Our own government recognized the value of this plan when it established the system of coinage.

The principle lies in fixing a unit, such as a dol p. 145 lar, or a pound, or a foot, and then making all divisions, or addition, in multiples of ten. Thus, we have one mill; ten mills to make a dime; ten dimes to make a dollar, and so on.

Basis of Measurement.—The question arose, what to use as the basis of measurement, and it was proposed to use the earth itself, as the measure. For this purpose the meridian line running around the earth at the latitude of Paris was selected.

One-quarter of this measurement around the globe was found to be 393,707,900 inches, and this was divided into 10,000,000 parts. Each part, therefore, was a little over 39.37 inches in length, and this was called a meter, which means *measure*.

A decimeter is one-tenth of that, namely, 3.937 inches; and a decameter 39.37, or ten times the meter, and so on.

For convenience the metrical table is given, showing lengths in feet and inches, in which only three decimal points are used.

Metrical Table, showing measurements in feet and inches: p. 146

METRICAL TABLE, SHOWING MEASUREMENTS IN FEET AND INCHES

Length	Inches	Feet
Millimeter	0.039	0.003
Centimeter	0.393	0.032
Decimeter	3.937	0.328
Meter	39.370	3.280
Decameter	393.707	32.808
Hectometer	3937.079	328.089
Kilometer	39370.790	3280.899
Myriameter	393707.900	32808.992

METRIC SYSTEM, SHOWING THE EQUIVALENTS IN OUR MEASURES

1 Myriameter	= 5.4 nautical miles, or 6.21 statute miles.
1 Kilometer	= 0.621 statute mile, or nearly 5/8 mile.

1 Hectometer	= 109.4 yards.
1 Decameter	= 0.497 chain, 1.988 rods.
1 Meter	= 39.37 inches, or nearly 3 ft. 3 3/8 inches.
1 Decimeter	= 3.937 inches.
1 Centimeter	= 0.3937 inch.
1 Millimeter	= 0.03937 inch.
1 Micron	= 1/25400 inch.
1 Hectare	= 2.471 acres.
1 Arc	= 119.6 square yards.
1 Centaire, or square meter	= 10.764 square feet. p. 147
1 Decastere	= 13 cubic yards, or about 2 3/4 cords.
1 Stere, or cubic meter	= 1.308 cubic yards, or 35.3 cubic feet.
1 Decistere	= 3 1/2 cubic feet.
1 Kiloliter	= 1 ton, 12 gal., 2 pints, 2 gills old wine measure.
1 Hectoliter	= 22.01 Imperial gals., or 26.4 U. S. gals.
1 Decaliter	= 2 gallons, 1 pint, 2 2/5 gills, imperial measure, or 2 gals., 2 qts., 1 pt., 1/2 gill, U. S.
1 Liter	= 1 pint, 3 gills, imperial, or 1 qt., 1/2 gill U. S. measure.
1 Decileter	= 0.704 gill, imperial, or 0.845 gill U. S. measure.
1 Millier	= 2,204.6 pounds avoirdupois.
1 Metric quintal	= 2 hundredweight, less 3 1/2 pounds, or 220 pounds, 7 ounces.
1 Kilogram	= 2 pounds, 3 ounces, 4 3/8 drams avoirdupois.
1 Hectogram	= 3 ounces, 8 3/8 drams avoirdupois.
1 Decagram	= 154.32 grains Troy.
1 Gram	= 15.432 grains.
1 Decigram	= 1.542 grain.
1 Centigram	= 0.154 grain.
1 Milligram	= 0.015 grain.

p. 148

CHAPTER XIII ToC

USEFUL INFORMATION FOR THE WORKSHOP

To find the circumference of a circle: Multiply the diameter by 3.1416.

To find the diameter of a circle: Multiply the circle by .31831.

To find the area of a circle: Multiply the square of the diameter by .7854.

To find the area of a triangle: Multiply the base by one-half the perpendicular height.

To find the surface of a ball: Multiply the square of the diameter by 3.1416.

To find the solidity of a sphere: Multiply the cube of the diameter by .5236.

To find the cubic contents of a cone: Multiply the area of the base by one-third the altitude.

Doubling the diameter of a pipe increases its capacity four times.

To find the pressure in pounds per square inch of a column of water: Multiply the height of the column in feet by .434.

Standard Horse-power: The evaporation of 30 pounds of water per hour from a feed water temperature of 1,000 degrees Fahrenheit into steam at 70 pounds gauge pressure

p. 149

To find the capacity of any tank in gallons: Square the diameter in inches, multiply by the length, and then by .0034.

In making patterns for aluminum castings provision must be made for shrinkage to a greater extent than with any other metal or alloy.

The toughness of aluminum can be increased by adding a small per cent. of phosphorus.

All alloys of metals having mercury are called *amalgams*.

A sheet of zinc suspended in the water of a boiler will produce an electrolytic action and prevent scaling to a considerable extent.

Hydrofluoric acid will not affect a pure diamond, but will dissolve all imitations.

A strong solution of alum put into glue will make it insoluble in water.

A grindstone with one side harder than the other can have its flinty side softened by immersing that part in boiled linseed oil.

One barrel contains 3 3/4 cubic feet.

One cubic yard contains 7 barrels.

To find the speed of a driven pulley of a given diameter: Multiply the diameter of the driving pulley by its speed or number of revolutions. Divide this by the diameter of the driven pulley. The result will be the number of revolutions of the driven pulley

p. 150

To find the diameter of a driven pulley that shall make any given number of revolutions in the same time: Multiply the diameter of the driving pulley by its number of revolutions, and divide the product by the number of revolutions of the driven pulley.

A piece of the well-known tar soap held against the inside of a belt while running will prevent it from slipping, and will not injure the belt.

Boiler scale is composed of the carbonate or the sulphate of lime. To prevent the formation it is necessary to use some substance which will precipitate these elements in the water. The cheapest and most universally used for this purpose are soda ash and caustic soda.

Gold bronze is merely a mixture of equal parts of oxide of tin and sulphur. To unite them they are heated for some time in an earthen retort.

Rusted utensils may be cleaned of rust by applying either turpentine or kerosene oil, and allowing them to stand over night, when the excess may be wiped off. Clean afterwards with fine emery cloth.

Plaster of paris is valuable for many purposes in a machine shop, but the disadvantage in handling it is, that it sets so quickly, and its use is, therefore, very much limited. To prevent quick setting mix a small amount of arrow root powder with p. 151 the plaster before it is mixed, and this will keep it soft for some time, and also increase its hardness when it sets.

For measuring purposes a tablespoon holds 1/2 ounce; a dessert-spoon 1/4 ounce; a teaspoon 1/8 ounce; a teacupful of sugar weighs 1/2 pound; two teacupsful of butter weigh 1 pound; 1 1/3 pints of powdered sugar weigh 1 pound; one pint of distilled water weighs 1 pound.

Ordinarily, 450 drops of liquid are equal to 1 ounce; this varies with different liquids, some being thicker in consistency than others, but for those of the consistency of water the measure given is fairly accurate

p. 152

CHAPTER XIV ToC

THE SIMPLICITY OF GREAT INVENTIONS, AND OF NATURE'S MANIFESTATIONS

If there is anything in the realm of mechanics which excites the wonder and admiration of man, it is the knowledge that the greatest inventions are the simplest, and that the inventor must take advantage of one law in nature which is universal in its application, and that is vibration.

There is a key to every secret in nature's great storehouse. It is not a complicated one, containing a multiplicity of wards and peculiar angles and recesses. It is the very simplicity in most of the problems which long served as a bar to discovery in many of the arts. So extremely simple have been some of the keys that many inventions resulted from accidents.

Invention Precedes Science.—Occasionally inventions were brought about by persistency and energy, and ofttimes by theorizing; but science rarely ever aids invention. The latter usually precedes science. Thus, reasoning could not show how it might be possible for steam to force water into a boiler against its own pressure. But the injector does this

p. 153

If, prior to 1876, it had been suggested that a sonorous vibration could be converted into an electrical pulsation, and transformed back again to a sonorous vibration, science would have proclaimed it impossible; but the telephone does it. Invention shows how things are done, and science afterwards explains the phenomena and formulates theories and laws which become serviceable to others in the arts.

Simplicity in Inventions.—But let us see how exceedingly simple are some of the great discoveries of man.

The Telegraph.—The telegraph is nothing but a magnet at each end of a wire, with a lever for an armature, which opens and closes the circuit that passes through the magnets and armature, so that an impulse on the lever, or armature, at one end, by making and breaking the circuit, also makes and breaks the circuit at the other end.

Telephone.—The telephone has merely a disk close to but not touching the end of a magnet. The sonorous vibration of the voice oscillates the diaphragm, and as the diaphragm is in the magnetic field of the magnet, it varies the pressure, so called, causing the diaphragm at the other end of the wire to vibrate in unison and give out the same sound originally imparted to the other diaphragm.

Transmitter.—The transmitter is merely a sen p. 154 sitized instrument. It depends solely on the principle of light contact points in an electric circuit, whereby the vibrations of the voice are augmented.

Phonograph.—The phonograph is not an electrical instrument. It has a diaphragm provided centrally with a blunt pin, or stylus. To make the record, some soft or plastic material, like wax, or tinfoil, is caused to move along so that the point of the stylus makes impressions in it, and the vibrations of the diaphragm cause the point to traverse a groove of greater or smaller indentations. When this groove is again presented to the stylus the diaphragm is vibrated and gives forth the sounds originally imparted to it when the indentations were made.

Wireless Telegraphy.—Wireless telegraphy depends for its action on what is called induction. Through this property a current is made of a high electro-motive force, which means of a high voltage, and this disturbs the ether with such intensity that the waves are sent out in all directions to immense distances.

The great discovery has been to find a mechanism sensitive enough to detect the induction waves. The instrument for this purpose is called a coherer, in which small particles cohere through the action of the electric waves, and are caused to fall apart mechanically, during the electrical impulses

p. 155

Printing Telegraph.—The printing telegraph requires the synchronous turning of two wheels. This means that two wheels at opposite ends of a wire must be made to turn at exactly the same rate of speed. Originally, this was tried by clock work, but without success commercially, for the reason that a pendulum does not beat with the same speed at the equator, as at different latitudes, nor at altitudes; and temperature also affects the rate. The solution was found by making the two wheels move by means of a timing fork, which vibrates with the same speed everywhere, and under all conditions.

Electric Motor.—The direct current electric motor depends for its action on the principle that likes repel, and unlikes attract. The commutator so arranges the poles that at the proper points, in the revolution of the armature, the poles are always presented to each other in such a way that as they approach each other, they are opposites, and thus attract, and as they recede from each other they repel. A dynamo is exactly the same, except that the commutator reverses the operation and makes the poles alike as they approach each other, and unlike as they recede.

Steel is simply iron, to which has been added a small per cent of carbon.

Quinine is efficient in its natural state, but it has p. 156 been made infinitely more effectual by the breaking up or changing of the molecules with acids. Sulphate of quinine is made by the use of sulphuric acid as a solvent.

Explosions.—Explosions depend on oxygen. While this element does not burn, a certain amount of it must be present to support combustion. Thus, the most inflammable gas or liquid will not burn or explode unless oxygenized. Explosives are made by using a sufficient amount, in a concentrated form, which is added to the fuel, so that when it is ignited there is a sufficient amount of oxygen present to support combustion, hence the rapid explosion which follows.

Vibration in Nature.—The physical meaning of vibration is best illustrated by the movement of a pendulum. All agitation is vibration. All force manifests itself in this way.

The painful brilliancy of the sun is produced by the rapid vibrations of the rays; the twinkle of the distant star, the waves of the ocean when ruffled by the winds; the shimmer of the moon on its crested surface; the brain in thinking; the mouth in talking; the beating of the heart; all, alike, obey the one grand and universal law of vibratory motion.

Qualities of Sound.—Sound is nothing but a succession of vibrations of greater or less magnitude. Pitch is produced by the number of vibra p. 157 tions; intensity by their force; and quality by the character of the article vibrated.

Since the great telephone controversy which took place some years ago there has been a wonderful development in the knowledge of acoustics, or sounds. It was shown that the slightest sound would immediately set into vibration every article of furniture in a room, and very sensitive instruments have been devised to register the force and quality.

The Photographer's Plate.—It is known that the chemical action of an object on a photographer's plate is due to vibration; each represents a force of different intensity, hence the varying shades produced. Owing to the different rates of vibrations caused by the different colors, the difficulty has been to photograph them, but this has now been accomplished. Harmony, or "being in tune," as is the common expression, is as necessary in light, as in music.

Some chemicals will bring out or "develop," the pictures; others will not. Colors are now photographed because invention and science have found the harmonizing chemicals.

Quadruplex Telegraphy.—One of the most remarkable of all the wonders of our age is what is known as duplex and quadruplex telegraphy. Every atom and impulse in electricity is oscillation. p. 158 The current which transmits a telegram is designated in the science as "vibratory."

But how is it possible to transmit two or more messages over one wire at the same time? It is by bringing into play the harmony of sounds. One message is sent in one direction in the key of A; another message in the other direction in B; and so any number may be sent, because the electrical vibrations may be tuned, just like the strings of a violin.

Electric Harmony.—Every sound produces a corresponding vibration in surrounding objects. While each vibrates, or is capable of transmitting a sound given to it by its vibratory powers, it may not vibrate in harmony.

When a certain key of a piano is struck every key has a certain vibration, and if we could separate it from the other sounds, it would reflect the same sound as the string struck, just the same as the walls of a room or the air itself would convey that sound.

But as no two strings in the instrument vibrate the same number of times each second, the rapid movement of successive sounds of the keys do not interfere with each other. If, however, there are several pianos in a room, and all are tuned the same pitch, the striking of a key on one instrument will p. 159 instantly set in vibration the corresponding strings in all the other instruments.

This is one reason why a piano tested in a music wareroom has always a more beautiful and richer sound than when in a drawing-room or hall, since each string is vibrated by the other instrument.

If a small piece of paper is balanced upon the strings of a violin, every key of the piano may be struck, except the one in tune, without affecting the paper; but the moment the same key is struck the vibration of the harmonizing pitch will unbalance the paper.

The musical sound of C produces 528 vibrations per second; D 616, and so on. The octave above has double the number of vibrations of the lower note. It will thus be understood why discord in music is not pleasant to the ear, as the vibrations are not in the proper multiples.

Odors.—So with odors. The sense of smell is merely the force set in motion by the vibration of the elements. An instrument called the *odophone* demonstrates that a scale or gamut exists in flowers; that sharp smells indicate high tones and heavy smells low tones. Over fifty odors have thus been analyzed.

The treble clef, note E, 4th space, is orange; note D, 1st space below, violet; note F, 4th space above clef, ambergris. To make a proper bouquet, there p. 160 fore the different odors must be harmonized, just the same as the notes of a musical chord are selected.

A Bouquet of Vibrations.—The odophone shows that santal, geranium, orange flower and camphor, make a bouquet in the key of

C. It is easy to conceive that a beautiful bouquet means nothing more than an agreeable vibratory sensation of the olfactory nerves.

Taste.—So with the sense of taste. The tongue is covered with minute cells surrounded by nervous filaments which are set in motion whenever any substance is brought into contact with the surface. Tasting is merely the movement of these filaments, of greater or less rapidity.

If an article is tasteless, it means that these filaments do not vibrate. These vibrations are of two kinds. They may move faster or slower, or they may move in a peculiar way. A sharp acute taste means that the vibrations are very rapid; a mild taste, slow vibrations.

When a pleasant taste is detected, it is only because the filaments are set into an agreeable motion. The vibrations in the tongue may become so rapid that it will be painful, just as a shriek becomes piercing to the ear, or an intense light dazzling to the eye; all proceed from the same physical force acting on the brain

p. 161

Color.—Color, that seemingly unexplainable force, becomes a simple thing when the principles of vibration are applied, and this has been fully explained by the spectroscope and its operation.

When the boy once appreciates that this force, or this motion in nature is just as simple as the great inventions which have grown out of this manifestation, he will understand that a knowledge of these things will enable him to utilize the energy in a proper way

p. 162

CHAPTER XV ToC

WORKSHOP RECIPES AND FORMULAS

In a work of this kind, dealing with the various elements, the boy should have at hand recipes or formulas for everything which comes within the province of his experiments. The following are most carefully selected, the objects being to present those which are the more easily compounded.

Adhesives for Various Uses.—Waterproof glue. Use a good quality of glue, and dissolve it in warm water, then add one pound of linseed oil to eight pounds of the glue. Add three ounces of nitric acid.

Leather or Card-board Glue. After dissolving good glue in water, to which a little turpentine has been added, mix it with a thick paste of starch, the proportion of starch to glue being about two to every part of glue used. The mixture is used cold.

A fine Belt Glue. Dissolve 50 ounces of gelatine in water, and heat after pouring off the excess water. Then stir in five ounces of glycerine, ten ounces of turpentine, and five ounces of linseed oil varnish. If too thick add water to suit.

For cementing Iron to Marble. Use 30 parts of Plaster of Paris, 10 parts of iron filings, and one p. 163 half part of sal ammoniac. These are mixed up with vinegar to make a fluid paste.

To cement Glass to Iron. Use 3 ounces of boiled linseed oil and 1 part of copal varnish, and into this put 2 ounces of litharge and 1 ounce of white lead and thoroughly mingle so as to make a smooth paste.

Water-proof Cement. Boiled linseed oil, 6 ounces; copal, 6 ounces; litharge, 2 ounces; and white lead, 16 ounces. To be thoroughly incorporated.

To unite rubber or leather to hard substances. One ounce of pulverized gum shellac dissolved in 9 1/2 ounces of strong ammonia, will make an elastic cement. Must be kept tightly corked.

For uniting iron to iron. Use equal parts of boiled oil, white lead, pipe clay and black oxide of manganese, and form it into a paste.

Transparent Cement. Unite 1 ounce of india rubber, 67 ounces of chloroform, and 40 ounces of mastic. This is to be kept together for a week, and stirred at times, when it will be ready for use.

To Attach Cloth to Metal. Water 100 parts, sugar 10 parts, starch 20 parts, and zinc chloride 1 part. This must be first stirred and made free of lumps, and then heated until it thickens.

United States Government Gum. Dissolve 1 part of gum arabic in water and add 4 parts of sugar p. 164 and 1 part of starch. This is then boiled for a few minutes, and thinned down as required.

To Make Different Alloys.—Silver-aluminum. Silver one-fourth part, and aluminum three-fourth parts.

Bell-metal. Copper, 80 parts; tin, 20 parts. Or, copper, 72 parts; tin, 26 parts; zinc, 2 parts. Or, copper 2; 1 of tin.

Brass. Copper, 66 parts; zinc, 32 parts; tin, 1 part; lead, 1 part.

Bronzes. Copper, 65 parts; zinc, 30 parts; tin, 5 parts. Or, copper, 85 parts; zinc, 10 parts; tin, 3 parts; lead, 2 parts.

German Silver. 52 parts of copper; 26 parts zinc; 22 parts nickel.

For Coating Mirrors. Tin, 70 parts; mercury, 30 parts.

Boiler Compounds.—To prevent scaling. Use common washing soda, or Glauber salts.

To Dissolve Celluloid.—Use 50 parts of alcohol and 5 parts of camphor for every 5 parts of celluloid. When the celluloid is put into the solution it will dissolve it.

To Soften Celluloid. This may be done by simply heating, so it will bend, and by putting it in steam, it can be worked like dough.

Clay Mixture for Forges.—Mix dry 20 parts of fire clay, 20 parts cast-iron turnings, one part p. 165 of common salt, and 1/2 part sal ammoniac, and then add water while stirring, so as to form a mortar of the proper consistency. The mixture will become very hard when heat is applied.

A Modeling Clay. This is made by mixing the clay with glycerine and afterwards adding vaseline. If too much vaseline is added it becomes too soft.

Fluids for Cleaning Clothes, Furniture, Etc.—For Delicate Fabrics. Make strong decoction of soap bark, and put into alcohol.

Non-inflammable Cleaner. Equal parts of acetone, ammonia and diluted alcohol.

Taking dried paint from clothing. Shake up 2 parts of ammonia water with 1 part of spirits of turpentine.

Cleaning Furniture, etc. Unite 2.4 parts of wax; 9.4 parts of oil of turpentine; 42 parts acetic acid; 42 parts citric acid; 42 parts white soap. This must be well mingled before using.

Removing Rust from Iron or Steel. Rub the surface with oil of tartar. Or, apply turpentine or kerosene, and after allowing to stand over night, clean with emery cloth.

For Removing Ink Stains from Silver. Use a paste made of chloride of lime and water.

To clean Silver-Plated Ware. Make a mixture of cream of tartar, 2 parts; levigated chalk, 2 p. 166 parts; and alum, 1 part. Grind up the alum and mix thoroughly.

Cleaning a Gas Stove. Make a solution of 9 parts of caustic soda and 150 parts of water, and put the separate parts of the stove in the solution for an hour or two. The parts will come out looking like new.

Cleaning Aluminum. A few drops of sulphuric acid in water will restore the luster to aluminum ware.

Oil Eradicator. Soap spirits, 100 parts; ammonia solution, 25; acetic ether, 15 parts.

Disinfectants.—Camphor, 1 ounce; carbolic acid (75 per cent.), 12 ounces; aqua ammonia, 10 drachms; soft salt water, 8 drachms.

Water-Closet Deodorant. Ferric chloride, 4 parts; zinc chloride, 5 parts; aluminum chloride, 4 parts; calcium chloride, 5 parts; magnesium chloride, 3 parts; and water sufficient to make 90 parts. When all is dissolved add to each gallon 10 grains of thymol and a quarter-ounce of rosemary that had been previously dissolved in six quarts of alcohol.

Odorless Disinfectants. Mercuric chloride, 1 part; cupric sulphate, 10 parts; zinc sulphate, 50 parts; sodium chloride, 65 parts; water to make 1,000 parts.

Emery for Lapping Purposes. Fill a pint bottle p. 167 with machine oil and emery flour, in the proportion of 7 parts oil and 1 part emery. Allow it to stand for twenty minutes, after shaking up well, then pour off half the contents, without disturbing the settlings, and the part so poured off contains only the finest of the emery particles, and is the only part which should be used on the lapping roller.

Explosives.—Common Gunpowder. Potassium nitrate, 75 parts; charcoal, 15 parts; sulphur, 10 parts.

Dynamite. 75 per cent. nitro-glycerine; 25 per cent. infusorial earth.

Giant Powder. 36 per cent. nitro-glycerine; 48 per cent. nitrate of potash; 8 per cent. of sulphur; 8 per cent. charcoal.

Fulminate. Chlorate of potassia, 6 parts; pure lampblack, 4 parts; sulphur, 1 part. A blow will cause it to explode.

Files.—How to Keep Clean. Olive oil is the proper substance to rub over files, as this will prevent the creases from filling up while in use, and preserve the file for a longer time, and also enable it to do better cutting.

To Renew Old Files. Use a potash bath for boiling them in, and afterwards brush them well so as to get the creases clean. Then stretch a cotton cloth between two supports, and after plunging the p. 168 file into nitric acid, use the stretched cloth to wipe off the acid. The object is to remove the acid from the ridges of the file, so the acid will only eat out or etch the deep portions between the ridges, and not affect the edges or teeth.

Fire Proof Materials or Substances.—For Wood. For the kind where it is desired to apply with a brush, use 100 parts sodium silicate; 50 parts of Spanish white, and 100 parts of glue. It must be applied hot.

Another good preparation is made as follows: Sodium silicate, 350 parts; asbestos, powdered, 350 parts; and boiling water 1,000 parts.

For Coating Steel, etc. Silica, 50 parts; plastic fire clay, 10 parts; ball clay, 3 parts. To be thoroughly mixed.

For Paper. Ammonium sulphate, 8 parts; boracic acid, 3 parts; borax, 2 parts; water, 100 parts. This is applied in a liquid state to the paper surface.

Floor Dressings.—Oil Stain. Neats' foot oil, 1 part; cottonseed oil, 1 part; petroleum oil, 1 part. This may be colored with anything desired, like burnt sienna, annatto, or other coloring material.

Ballroom Powder. Hard paraffine, 1 pound; powdered boric acid, 7 pounds; oil of lavender, 1 drachm; oil of neroli, 20 minims.

Foot Powders.—For Perspiring Feet. Balsam p. 169 Peru, 15 minims; formic acid, 1 drachm; chloral hydrate, 1 drachm; alcohol to make 3 ounces.

For Easing Feet. Tannaform, 1 drachm; talcum, 2 drachms; lycopodium, 30 grains.

Frost Bites. Carbolized water, 4 drachms; nitric acid, 1 drop; oil of geranium, 1 drop.

Glass.—To cut glass, hold it under water, and use a pair of shears.

To make a hole through glass, place a circle of moist earth on the glass, and form a hole in this the diameter wanted for the hole, and in this hole pour molten lead, and the part touched by the lead will fall out.

To Frost Glass. Cover it with a mixture of 6 ounces of magnesium sulphate, 2 ounces of dextrine, and 20 ounces of water. This produces a fine effect.

To imitate ground glass, use a composition of sandarac, 2 1/2 ounces; mastic, 1/2 ounce; ether, 24 ounces; and benzine, 16 ounces.

Iron and Steel.—How to distinguish them. Wash the metal and put it into a solution of bichromate of potash to which has been added a small amount of sulphuric acid. In a minute or so take out the metal, wash and wipe it. Soft steel and cast iron will have the appearance of an ash-gray tint; tempered steels will be black; and pud p. 170 dled or refined irons will be nearly white and have a metallic reflection.

To Harden Iron or Steel. If wrought iron, put in the charge 20 parts, by weight, of common salt, 2 parts of potassium cyanide, .3 part of potassium bichromate, .15 part of broken glass.

To harden cast iron, there should be added to the charge the following: To 60 parts of water, add 2 1/2 parts of vinegar, 3 parts of common salt, and .25 part of hydrochloric acid.

To soften castings: Heat them to a high temperature and cover them with fine coal dust and allow to cool gradually.

Lacquers.—For Aluminum. Dissolve 100 parts of gum lac in 300 parts of ammonia and heat for an hour moderately in a water bath. The aluminum must be well cleaned before applying. Heat the aluminum plate afterwards.

For Brass. Make a compound as follows; Annatto, 1/4 ounce; saffro, 1/4 ounce; turmeric, 1 ounce; seed lac, 3 ounces; and alcohol, 1 pint. Allow the mixture to stand for three days, then strain in the vessel which contains the seed lac, and allow to stand until all is dissolved.

For Copper. Heat fine, thickly liquid amber varnish so it can be readily applied to the copper, and this is allowed to dry. Then heat the coated object until it commences to smoke and turn brown

p. 171

Lubricants.—Heavy machinery oils. Use paraffine, 8 pounds; palm oil, 20 pounds; and oleonaptha, 12 pounds. Dissolve the paraffine in the oleonaptha at a temperature of 160 degrees and then stir in the palm oil a little at a time.

For Cutting Tools. Heat six gallons of water and put in three and a half pounds of soft soap and a half gallon of clean refuse oil. It should be well mixed.

For high-speed bearings. Use flaky graphite and kerosene oil. Apply this as soon as there is any indication of heating in the bearings.

For lathe centers, one part of graphite and four parts of tallow thoroughly mixed and applied will be very serviceable.

For Wooden Gears. Use tallow, 30 parts; palm oil; 20 parts; fish oil, 10 parts; and graphite, 20 parts.

Paper.—Fire Proof Paper.—Make the following solution: Ammonium sulphate, 8 parts; boracic acid, 3 parts; water, 100 parts. Mix at a temperature of 120 degrees. Paper coated with this will resist heat.

Filter Paper. Dip the paper into nitric acid of 1.433 specific gravity, and subsequently wash and dry it. This makes a fine filtering body.

Carbon Paper. A variety of substances may be used, such as fine soot or ivory black, ultramarine p. 172 or Paris blue. Mix either with fine grain soap, so it is of a uniform consistency and then apply to the paper with a stiff brush, rubbing it in until it is evenly spread over the surface.

Tracing Paper. Take unsized paper and apply a coat of varnish made of equal parts of Canada balsam and oil of turpentine. To increase the transparency give another coat. The sheets must be well dried before using.

Photography.—Developers.

1. Pure water, 30 ounces; sulphite soda, 5 ounces; carbonate soda, 2 1/2 ounces.

2. Pure water, 24 ounces; oxalic acid, 15 grains; pyrogallic acid, 1 ounce.

To develop use of solution 1, 1 ounce; solution 2, 1/2 ounce; and water, 3 ounces.

Stock solutions for developing: Make solution No. 1 as follows: water, 32 ounces; tolidol, I ounce; sodium sulphate, 1 1/2 ounces.

Solution No. 2: Water, 32 ounces; sodium sulphate.

Solution No. 3: Water, 32 ounces; sodium carbonate, from 4 to 6 ounces.

Fixing bath. Add two ounces of S. P. C. clarifier (acid bisulphate of sodium) solution to one quart of hypo solution 1 in 5.

Clearing solution. Saturated solution of alum, 20 ounces; and hydrochloric acid, 1 ounce. p. 173 Varnish. Brush over the negative a solution of equal parts of benzol and Japanese gold size.

Plasters.—Court Plaster. Use good quality silk, and on this spread a solution of isinglass warmed. Dry and repeat several times, then apply several coats of balsam of Peru. Or,

On muslin or silk properly stretched, apply a thin coating of smooth strained flour paste, and when dry several coats of colorless gelatine are added. The gelatine is applied warm, and cooled before the fabric is taken off.

Plating.—Bronze coating. For antiques, use vinegar, 1,000 parts; by weight, powdered bloodstone, 125 parts; plumbago, 25 parts. Apply with brush.

For brass where a copper surface is desired, make a rouge with a little chloride of platinum and water, and apply with a brush.

For gas fixtures. Use a bronze paint and mix with it five times its volume of spirit of turpentine, and to this mixture add dried slaked lime, about 40 grains to the pint. Agitate well and decant the clear liquid.

Coloring Metals.—Brilliant black for iron. Selenious acid, 6 parts; cupric sulphate, 10 parts; water 1,000 parts; nitric acid, 5 parts.

Blue-black. Selenious acid, 10 parts; nitric acid, 5 parts; cupric sulphate; water, 1,000 parts. The p. 174 colors will be varied dependent on the time the objects are immersed in the solution.

Brass may be colored brown by using an acid solution of nitrate of silver and bismuth; or a light bronze by an acid solution of nitrate of silver and copper; or black by a solution of nitrate of copper.

To copper plate aluminum, take 30 parts of sulphate of copper; 30 parts of cream of tartar; 25 parts of soda; and 1,000 parts of water. The article to be coated is merely dipped into the solution.

Polishers.—Floor Polish. Permanganate of potash in boiling water, applied to the floor hot, will produce a stain, the color being dependent on the number of coats. The floor may them be polished with beeswax and turpentine.

For Furniture. Make a paste of equal parts of plaster of paris, whiting, pumice stone and litharge, mixed with Japan dryer, boiled linseed oil and turpentine. This may be colored to suit. This will fill the cracks of the wood. Afterwards rub over the entire surface of the wood with a mixture of 1 part Japan, 2 of linseed oil, and three parts of turpentine, also colored, and after this has been allowed to slightly harden, rub it off, and within a day or two it will have hardened sufficiently so that the surface can be polished.

Stove Polish. Ceresine, 12 parts; Japan wax, p. 175 10 parts; turpentine oil, 100 parts; lampblack, 12 parts; graphite, 10 parts. Melt the ceresine and wax together, and cool off partly, and then add and stir in the graphite and lampblack which were previously mixed up with the turpentine.

Putty. — Black Putty. Whiting and antimony sulphide, and soluble glass. This can be polished finely after hardening.

Common Putty. Whiting and linseed oil mixed up to form a dough.

Rust Preventive. — For Machinery. Dissolve an ounce of camphor in one pound of melted lard. Mix with this enough fine black lead to give it an iron color. After it has been on for a day, rub off with a cloth.

For tools, yellow vaseline is the best substance.

For zinc, clean the plate by immersing in water that has a small amount of sulphuric acid in it. Then wash clean and coat with asphalt varnish.

Solders. — For aluminum. Use 5 parts of tin and 1 part of aluminum as the alloy, and solder with the iron or a blow pipe.

Yellow hard solder. Brass, 3 1/2 parts; and zinc, 1 part.

For easily fusing, make an alloy of equal parts of brass and zinc.

For a white hard solder use brass, 12 parts; zinc, 1 part; and tin, 2 parts.

p. 176

Soldering Fluxes. — For soft soldering, use a solution of chloride of zinc and sal ammoniac. Powdered rosin is also used.

For hard soldering, borax is used most frequently.

A mixture of equal parts of cryolite and barium chloride is very good in soldering bronze or aluminum alloys.

Other hard solders are alloyed as follows: brass, 4 parts; and zinc, 5 parts. Also brass, 7 parts; and zinc, 2 parts.

Steel Tempering-.-Heat the steel red hot and then plunge it into sealing wax.

For tempering small steel springs, they may be plunged into a fish oil which has a small amount of rosin and tallow.

Varnishes. — Black Varnish. Shellac, 5 parts; borax, 2 parts; glycerine, 2 parts; aniline black, 6 parts; water, 45 parts. Dissolve the shellac in hot water and add the other ingredients at a temperature of 200 degrees.

A good can varnish is made by dissolving 15 parts of shellac, and adding thereto 2 parts of Venice turpentine, 8 parts of sandarac, and 75 parts of spirits.

A varnish for tin and other small metal boxes is made of 75 parts alcohol, which dissolves 15 parts of shellac, and 3 parts of turpentine.

Sealing Wax.—For modeling purposes. White p. 177 wax, 20 parts; turpentine, 5 parts; sesame oil, 2 parts; vermilion, 2 parts.

Ordinary Sealing. 4 pounds of shellac, 1 pound Venice turpentine, add 3 pounds of vermilion. Unite by heat.

p. 178

HANDY TABLES

TABLE OF WEIGHTS FOR ROUND AND SQUARE STEEL.

The Estimate is on the basis of Lineal Feet. 1 cu. ft. of Steel — 490 lbs.

Sizes in Inches	Weight in Pounds Round	Square	Sizes in Inches	Weight in Pounds Round	Square
1/16	.110	.013	1 1/16	3.014	3.400
1/8	.042	.053	1 1/8	3.379	3.838
3/16	.094	.119	1 3/16	3.766	4.303
1/4	.167	.212	1 1/4	4.173	4.795
5/16	.261	.333	1 5/16	4.600	5.312
3/8	.375	.478	1 3/8	5.049	5.857
7/16	.511	.651	1 7/16	5.518	6.428
1/2	.667	.850	1 1/2	6.008	7.650
9/16	.845	1.026	1 9/16	6.520	7.650
5/8	1.043	1.328	1 5/8	7.051	8.301
11/16	1.262	1.608	1 11/16	7.604	8.978
3/4	1.502	1.913	1 3/4	8.178	10.410
13/16	1.773	2.245	1 13/16	8.773	11.170
7/8	2.044	2.603	1 7/8	9.388	11.950
15/16	2.347	2.989	1 15/16	10.020	12.760
1	2.670	3.400	2	10.680	13.600

p. 179

WEIGHT OF FLAT STEEL BARS.

Thickness   Width

in
Inches

1/16	.212	.265	.32	.372	.425	.477	.53	.588	.63
1/8	.425	.53	.64	.745	.85	.955	1.06	1.17	1.27
3/16	.638	.797	.957	1.11	1.28	1.44	1.59	1.75	1.91
1/4	.85	1.06	1.28	1.49	1.70	1.91	2.12	2.34	2.55
5/16	1.06	1.33	1.59	1.86	2.12	2.39	2.65	2.92	3.19
3/8	1.28	1.59	1.92	2.23	2.55	2.87	3.19	3.51	3.83
7/16	1.49	1.85	2.23	2.60	2.98	3.35	3.72	4.09	4.46
1/2	1.70	2.12	2.55	2.98	3.40	3.83	4.25	4.67	5.10
9/16	1.92	2.39	2.87	3.35	3.83	4.30	4.78	5.26	5.74
5/8	2.12	2.65	3.19	3.72	4.25	4.78	5.31	5.84	6.38
11/16	2.34	2.92	3.51	4.09	4.67	5.26	5.84	6.43	7.02
3/4	2.55	3.19	3.83	4.47	5.10	5.75	6.38	7.02	7.65
13/16	2.76	3.45	4.14	4.48	5.53	6.21	6.90	7.60	8.29
7/8	2.98	3.72	4.47	5.20	5.95	6.69	7.44	8.18	8.93
15/16	3.19	3.99	4.78	5.58	6.38	7.18	7.97	8.77	9.57
1	3.40	4.25	5.10	5.95	6.80	7.65	8.50	9.35	10.20

p. 180

AVOIRDUPOIS WEIGHT.

For Merchandise of all kinds.

16	Drams (dr.) make	1 Ounce (oz.)
16	Ounces make	1 Pound (pd.)
25	Pounds make	1 Quarter (qr.)
4	Quarters, or 100 lbs., make	1 Hundredweight (cwt.)
20	Hundredweights make	1 Ton (T.)
2,240	Pounds make	1 Long ton (L. T.)

TROY WEIGHT.

For Gold, Silver, and Precious Metals.

24	Grains (gr.) make	1 Pennyweight (pwt.)
20	Pennyweights make	1 Ounce (oz.)
12	Ounces make	1 Pound (pd.)

APOTHECARIES WEIGHT.

For Drugs, Medicals and Chemicals.

20	Grains (gr.) make	1 Scruple (sc.)
3	Scruples make	1 Dram (dr.)
8	Drams make	1 Pound (pd.)
12	Ounces make	1 Pound (pd.)

p. 181

LINEAR MEASURE.

For Surveyors' Use.

12	Inches make	1 Foot
3	Feet make	1 Yard
5 1/2	Yards make	1 Rod
40	Rods make	1 Furlong
8	Furlongs make	1 Mile

LONG MEASURE.

12	Inches make	1 Foot
3	Feet make	1 Yard
6	Feet make	1 Fathom
5 1/2	Yards make	1 Rod or pole
40	Poles make	1 Furlong
8	Furlongs make	1 Mile
3	Miles make	1 League
69 1/2	Leagues make	1 Degree

SQUARE MEASURE.

144	square inches make	1 square foot
9	square feet make	1 square yard
30 1/2	square yards make	1 square pole
40	square poles make	1 square rod
4	square rods make	1 acre
640	square acres make	1 acre mile
9	square feet make	1 square yard

p. 182

SOLID OR CUBIC MEASURE.

1,728	Cubic inches make	1 Cubic foot
27	Cubic feet make	1 Cubic yard
128	Cubic feet make	1 Cord of wood
24 3/4	Cubic feet make	1 Perch of stone

DRY MEASURE.

2	Pints make	1 Quart (qt.)
8	Quarts make	1 peck (pk.)
4	Pecks make	1 Bushel (bu.)
36	Bushels make	1 Chaldron (ch.)

LIQUID MEASURE.

4	Gills (g.) make	1 Pint (pt.)
4	Quarts make	1 Gallon (gal.)
31 1/2	Gallons make	1 Barrel (bbl.)
2	Bbls., or 63 gals., make	1 Hogshead (hhd.)

PAPER MEASURE.

24	Sheets (sh.) make	1 Quire (qu.)
20	Quires make	1 Ream (r.)
10	Reams make	1 Bale (ba.) or bundle.

p. 183

TABLE OF TEMPERATURES.

Greatest artificial cold	220	degrees	below	Fahr.
" natural "	39	"	"	"
Mercury freezes	73.7	"	"	"
Mixture of snow and salt	4	"	"	"
Greatest density of water at	39.2	"	above	"
Blood Heat	97.9	"	"	"
Alcohol boils	172.4	"	"	"
Water boils	212	"	"	"
Mercury boils	662	"	"	"
Sulphur boils	824	"	"	"
Silver melts	1,749	"	"	"
Cast iron melts	2,786	"	"	"

STRENGTH OF VARIOUS METALS.

The tests are made by using a cubic inch of the metal and compressing it, and by trying to draw apart a square inch of metal. Indicated in pounds

p. 184

	Tension	Compression
Aluminum	15,000	12,000
Brass, cast	24,000	30,000
Bronze, gun metal	32,000	20,000
" manganese	60,000	120,000
" phosphor	50,000	. . . . . .
Copper, cast	24,000	40,000
" wire annealed	36,000	. . . . . .
" unannealed	60,000	. . . . . .
Iron, cast	15,000	. . . . . .
" " annealed	60,000	80,000

" " unannealed	80,000	. . . . . .
" wrought	48,000	46,000
Lead, cast	2,000	. . . . . .
Steel castings	70,000	70,000
" plow	270,000	. . . . . .
" structural	60,000	60,000
" wire annealed	80,000	. . . . . .
" crucible	180,000	. . . . . .
Tin	3,800	6,000

p. 185

FREEZING MIXTURES

MIXTURES	Temperature Changes in Degrees Fahrenheit	
	From	To
Common salt, 1 part; snow, 3 parts	32	zero .0
Common salt, 1 part; snow 1 part	32	- .4
Calcium chloride, 3 parts; snow 1 part	32	-27
Calcium chloride, 2 parts; snow 1 part	32	-44
Sal ammoniac, 5 parts; salt-peter 5 parts; water 16 parts	50	-10
Sal ammoniac, 1 parts; salt-peter 1 part; water 1 part	46	-11
Ammonium nitrate, 1 part; water 1 part	50	- 3
Potassium hydrate, 4 parts; snow 3 parts	32	-35

IGNITION TEMPERATURES.

Phosphorus	120	degrees	Fahrenheit
Bi-sulphide of carbon	300	"	"

Gun-cotton	430	"	"
Nitro-glycerine	490	"	"
Phosphorus, amorphous	500	"	"
Rifle powder	550	"	"
Charcoal	660	"	"
Dry pine wood	800	"	"
Oak	900	"	"

POWER AND HEAT EQUIVALENTS.

In studying matters pertaining to power and heat, certain terms are used, such as horsepower, horsepower-hours, watts, watt-hours, kilowatt, kilowatt-hours, foot-pounds, joule, and B. T. U. (British Thermal Unit).

The following tables give a comprehensive idea of the values of the different terms:

1 Horsepower-hour	= 0.746 kilowatt-hour = 1,980,000 foot-pounds of water evaporated at 212 degrees Fahrenheit, raised from 62 degrees to 212 degrees.
1 Kilowatt-hour	= 1,000 watt-hours = 1.34 horse-power-hours = 2,653,200 foot-pounds = 3,600,000 joules = 3,420 B. T. U. = 3.54 pounds of water evaporated at 212 degrees = 22.8 pounds of water raised from 62 to 212 degrees.
1 Horsepower	= 746 watts = 0.746 kilowatts.= 33,000 foot-pounds per second = 2,550 B. T. U. per min. = 0.71 B. T. U. per second = 2.64 pounds of water evaporated per hour at 212 degrees.
1 Kilowatt	= 1,000 watts = 1.34 horsepower = 2,653,200 foot-pounds per hour = 44,220 foot-pounds per min. = 737 foot-pounds per second = 3,420 B. T. U. per hour = 57 B. T. U. per min. = 0.95 B. T. U. per second = 3.54 pounds of water p. 187 evaporated per hour at 212.
1 Watt	= 1 joule per second = 0.00134 horse-power = 0.001 kilowatt = 342 B. T. U. per hour = 44.22 foot-pounds per min. = 0.74 foot-pounds per second = 0.0035 pounds of water evaporated per hour at 212 degrees.
1 B. T. U. (British Thermal Unit)	= 1,052 watt-seconds = 778 foot-pounds = 0.252 calorie = 0.000292 kilowatt-hours = 0.000391 horse-power-hour = 0.00104 pounds of water evaporated at 212 degrees.

1 Foot-pound	= 1.36 joule = 0.000000377 kilowatt-hour = 0.00129 B. T. U. = 0.0000005 horsepower-hour.
1 Joule	= 1 watt-second = 0.000000278 kilowatt-hour = 0.00095 B. T. U. = 0.74 foot-pounds.

p. 188

CHAPTER XVII ToC

INVENTIONS AND PATENTS, AND INFORMATION ABOUT THE RIGHTS AND DUTIES OF INVENTORS AND WORKMEN

There is no trade or occupation which calls forth the inventive faculty to a greater degree than the machinist's. Whether it be in the direction of making some new tool, needed in some special work, or in devising a particular movement, or mechanical expedient, the machinist must be prepared to meet the issues and decide on the best structural arrangement.

Opportunities also come daily to the workers in machine shops to a greater extent than other artisans, because inventors in every line bring inventions to them to be built and experimentally tested.

A knowledge of the rights and duties of inventors, and of the men who build the models, is very desirable; and for your convenience we append the following information:

The inventor of a device is he who has conceived an idea and has put it into some concrete form.

A mere idea is not an invention.

The article so conceived and constructed, must p. 189 be both *new* and *useful*. There must be some utility. It may be simply a toy, or something to amuse.

If A has an idea, and he employs and pays B to work out the device, and put it into practical shape, A is the inventor, although B may have materially modified, or even wholly changed it. B is simply the agent or tool to bring it to perfection, and his pay for doing the work is his compensation.

An inventor has two years' time within which he may apply for a patent, after he has completed his device and begun the sale of it. If he sells the article for more than two years before applying for a patent, this will bar a grant.

Two or more inventors may apply for a patent, provided each has contributed something toward bringing it to its perfected state. Each cannot apply separately. The patent issued will be owned by them jointly.

Joint owners of a patent are not partners, unless they have signed partnership papers respecting the patent. Because they are partners in some other enterprise, disconnected from the patent, that does not constitute them partners in the patent. They are merely joint owners.

If they have no special agreement with respect to the patent each can grant licenses to manufac p. 190 ture, independently of the others, without being compelled to account to the others, and each has a right to sell his interest without asking permission of the others.

An *inventor* is one who has devised an invention. A *patentee* is one who owns a patent, or an interest in one, be he the inventor or not.

The United States government does not grant Caveats. The only protection offered is by way of patent.

A patent runs for a period of seventeen years, and may be renewed by act of Congress only, for a further term of seven years.

An interference is a proceeding in the Patent Office to determine who is the first inventor of a device. The following is a brief statement of the course followed:

When two or more applicants have applications pending, which, in the opinion of the Examiner, appear to be similar, the Office may declare an interference.

If an applicant has an application pending, and the Examiner rejects it on reference to a patent already issued, the applicant may demand an interference, and the Office will then grant a hearing to determine which of the two is entitled to the patent.

The first step, after the declaration of interfer p. 191 ence, is to request that each applicant file a preliminary statement, under oath, in which he must set forth the following:

First: The date of conception of the invention.

Second: Date of the first reduction to writing, or the preparation of drawings.

Third: Date of making of the first model or device.

Fourth: When a complete machine was first produced.

These statements are filed in the Patent Office, and opened on the same day, and times are then set for the respective parties to take testimony.

If one of the parties was the first to conceive and reduce to practice, as well as the first to file his application, he will be adjudged to be the first inventor, without necessitating the taking of testimony.

If, on the other hand, one was the first to conceive, and the other the first to file, then testimony will be required to determine the question of invention.

The granting of a patent is not conclusive that the patentee was, in reality, the first inventor. The law is that the patent must issue to the *first* inventor, and if it can be proven that another party was the first, a new patent will issue to the one who thus establishes his right. The Commis p. 192 sioner of Patents has no right to take away the patent first issued. Only the Courts are competent to do this.

A patent is granted for the right to *make*, to *use* and to *vend*.

An owner of a patent cannot sell the right only to make, or to sell, or to use. Such a document would be a simple license, only, for that particular purpose.

A patent may be sold giving a divided, or an undivided right.

A divided right is where a State, or any other particular territorial right is granted. An undivided right is a quarter, or a half, or some other portion in the patent itself.

If an inventor assigns his invention, and states in the granting clause that he conveys "all his right and title in and to the invention," or words to that effect, he conveys all his rights throughout the world.

If the conveyance says, "all rights and title in and throughout the United States," he thereby reserves all other countries.

If a patent is issued, and the number and date of the patent are given, the assignment conveys the patent for the United States only, unless foreign countries are specifically mentioned.

To convey an invention or patent, some definite p. 193 number or filing date must be given in the document, with sufficient clearness and certainty to show the intent of the assignor.

An invention does not depend on quantity, but on quality. It is that which produces a new and a useful *result*.

In the United States patents are granted for the purpose of promoting the useful arts and sciences.

In England, and in many other foreign countries, patents are granted, not on account of any merit on the part of the inventor, but as a favor of the crown, or sovereign.

Originally patents were granted by the crown for the exclusive privilege in dealing in any commodity, and for this right a royal fee was exacted. From this fact the term *royalty* originated.

An international agreement is now in force among nearly all countries, which respects the filing of an application in any country, for a period of one year in the other countries.

In making an application for a patent, a petition is required, a specification showing its object, use, and particular construction, followed by a claim, or claims, and accompanied by a drawing, if the invention will permit of it, (which must be made in black, with India ink), and an oath.

The oath requires the following assertions: That the applicant is the first and original inventor p. 194 of the device, and that he does not know and does not believe the same was ever known or used before his invention or more than two years before his application.

He must also further allege that the invention was not patented or described in any printed publication here or abroad, and not manufactured more than two years prior to the application, and that he has not made an application, nor authorized any one to do so more than two years prior to his application.

The first Government fee is $15, payable at the time of filing, and the second and final fee is $20, payable at the time the patent is ordered to issue.

The filing of an application for patent is a secret act, and the Patent Office will not give any information to others concerning it, prior to the issue of the patent

GLOSSARY OF WORDS
USED IN TEXT OF THIS VOLUME ToC

Abrupt.	Suddenly; coming without warning.
Abrasive.	A material which wears away.
Actuate.	Influenced, as by sudden motive; incited to action.
Accumulate.	To bring together; to amass; to collect.
Acoustics.	The branch of physics which treats of sound.
Adhesion.	To hold together; a molecular force by means of which particles stick together.
Affinity.	Any natural drawing together; the property or force in chemicals to move toward each other.
Aggravate.	To incite; to make worse or more burdensome.
Alloy.	A combination of two or more metals.
Altitude.	Height; a vertical distance above any point.
Alkali.	Any substance which will neutralize an acid, as lime, magnesia, and the like.
Amalgam.	Any compound of metal which has mercury as one of the elements.
Amiss.	Wrong, fault, misdeed.
Annealing.	A process of gradually heating and cooling metals, whereby hardness and toughness are brought about.
Angle plate.	A metal structure which has two bodies, or limbs, at right angles to each other.
Analysis.	The separating of substances into their elementary forms.
Anchor bolt.	A structure intended to be placed in a hole in a wall, and held there by a brew which expands a part of the structure. p. 196
Apprentice.	One who is learning a trade or occupation.

Artificial.	That which resembles the original; made in imitation of.
Arbor.	A shaft, spindle, mandrel, or axle.
Armature.	A metallic body within the magnetic field of a magnet.
Arbitrary.	Stubborn determination. Doing a thing without regard to consequences.
Artisan.	One skilled in any mechanical art.
Attributable.	That which belongs to or is associated with.
Automatically.	Operating by its own structure, or without outside aid.
Augmented.	Added to; to increase.
Auxiliary.	To aid; giving or furnishing aid.
Avoirdupois.	The system of weights, of which the unit is sixteen ounces.
Back-saw.	A saw which has a rib at its upper margin.
Barleycorn.	A grain of barley.
Bastard.	A coarse-grained file.
B. T. U.	British Thermal Unit.
Back-gear.	That gear on a lathe for changing the feed.
Bevel.	Not in a right line; slanting; oblique.
Bibb.	A form of water faucet.
Bit, or bitt.	A form of tool for cutting purposes on a lathe, planer, shaper, or drilling machine.
Borax.	A white crystalline compound, of a sweetish taste. Chemically it is sodium biborate.
Buffs.	Usually a wheel covered with leather or cloth, and having emery dust on it, for fine polishing purposes.
Buffeted.	Thrown back.
Bronze.	An alloy of copper and tin.
Calcium.	Lime.
Cant.	A form of lever. p. 197

Carbonate.	A salt of carbonic acid.
Caustic.	Capable of corroding or eating away.
Capillary.	That quality of a liquid which causes it to move upwardly or along a solid with which it is in contact.
Caliper.	An instrument for spanning inside and outside dimensions.
Centripetal.	The force which tends to draw inwardly, or to the center.
Centrifugal.	The outwardly-moving force from a body.
Centering.	To form a point equidistant from a circular line.
Chloride.	A compound of chlorine with one or more positive elements, such as, for instance, salt.
Circular pitch.	The measurement around a gear taken at a point midway between the base and end of the teeth.
Circumference.	The outside of a circular body.
Clef.	A character placed on a staff of music to determine the pitch.
Clutch.	A mechanical element for attaching one part to another.
Chuck, Independent.	A disk of metal to be attached to the live spindle of a lathe, and which has on its face a set of dogs which move radially independently of each other.
Chuck, Universal.	A disk to be attached as above, provided with dogs which are connected so they move radially in unison with each other.
Classified.	Arranged in order, in such a manner that each of a kind is placed under a suitable heading.
Clearance.	To provide a space behind the cutting edge of a tool which will not touch the work being cut.
Consistency.	Harmonious; not contradictory.
Coherer.	That instrument in a wireless telegraphy apparatus which detects the electrical impulses. p. 198
Commutator.	The cylindrical structure on the end of an arma-

	ture, which is designed to change the polarity of the current.
Concentrated.	Brought together at one point.
Coinage.	The system of making money from metals.
Compound.	The unity of two or more elements.
Constant.	Being insistent and consistent; also a term to be used in a problem which never varies.
Conversion.	The change from one state to another.
Cone.	A body larger at one end than at another; usually applied to a form which is cylindrical in shape but tapering, from end to end.
Compression.	The bringing together of particles, or molecules.
Convolute.	A spiral form of winding, like a watch spring.
Coiled.	A form of winding, like a string wound around a bobbin.
Conductivity.	Applied generally to the quality of material which will carry a current of electricity; also a quality of a material to convey heat.
Cohesion.	The force by which the molecules of the same kind are held together.
Concentric.	A line which is equidistant at all points from a center.
Confined.	Held within certain bounds.
Corpuscular.	Molecular or atomic form.
Converge.	To come together from all points.
Concave.	A surface which is depressed or sunken.
Convex.	A surface which is raised, or projects beyond the surface of the edges.
Component.	One of the elements in a problem or in a compound.
Coefficient.	A number indicating the degree or quality possessed by a substance. An invariable unit.
Cube.	A body having six equal sides. p. 199

Cross-section.	A term used to designate that line which is at right angles to the line running from the view point.
Cross slide.	The metal plate on a lathe which holds the tool post, and which is controlled, usually, by a screw.
Contiguous.	Close to; near at hand.
Countersink.	The depression around a bore.
Collet.	A collar, clutch or clamping piece, which has jaws to hold a bar or rod.
Countershaft.	A shaft which has thereon pulleys or gears to connect operatively with the gears or pulleys on a machine, and change the speed.
Conducive.	Tending to; promotive of a result.
Corundum.	An extremely hard aluminum oxide used for polishing.
Cold chisel.	A term applied to an extremely hard chisel used for cutting and chipping metal.
Combustion.	The action or operation of burning.
Conjunctively.	Acting together.
Comparatively.	Similitude or resemblance, one with another.
Cotter.	A key to prevent a wheel turning on its shaft.
Dead center.	A term used to designate the inoperative point of the crank.
Depicting.	Showing; setting forth.
Deodorant.	A substance which will decompose odors.
Developer.	A chemical which will bring out the picture in making the film or plate in photography.
Decimeter.	The length of one-tenth of a meter in the metric system.
Decameter.	The length of ten meters in the metric system.
Defective.	Not perfect; wrong in some particular.
Diaphragm.	A plate, such as used in a telephone system, to receive and transmit vibrations. p. 200

| Dissolving. | To change from a solid to a liquid condition. |

Dissolving. To change from a solid to a liquid condition.

Division plate. A perforated plate in a gear-cutting machine, to aid in dividing the teeth of a wheel.

Dispelled. To drive away or scatter.

Disinfectant. A material which will destroy harmful germs.

Diametral pitch. The number of teeth in a gear as calculated on the pitch line.

Dimension. Measurement; size.

Ductility. That property of metal which permits it to be drawn out, or worked.

Dividers. An instrument, like a compass, for stepping off measurements, or making circles.

Diverge. Spreading out from a common point.

Drift. A cutting tool for smoothing a hole in a piece of metal.

Duplex. Two; double.

Dynamite. An explosive composed of an absorbent, like earth, combined with nitro-glycerine.

Dynamometer. An instrument for measuring power developed.

Eccentric. Out of center.

Echoes. The reflection of sound.

Effervesce. The action due to the unity of two opposite chemicals.

Efficiency. The term applied to the quality of effectiveness.

Ellipse. A form which is oblong, or having a shape, more or less, like the longitudinal section of an egg.

Electrolytic. The action of a current of water passing through a liquid, and decomposing it, and carrying elements from one electrode to the other.

Elasticity. The quality in certain substances to be drawn out of their normal shape, and by virtue of which they will resume their original form when released. p. 201

Embedded.	To be placed within a body or substance.
Emerge.	To come out of.
Emphasize.	To lay particular stress upon.
Emery.	A hard substance, usually some of the finely divided precious stones, and used for polishing and grinding purposes.
Enormous.	A large amount; great in size.
Enunciated.	Proclaimed; given out.
Equalization.	To put on an even basis; to make the same comparatively.
Eradicator.	To take out; to cause to disappear.
Escapement.	A piece of mechanism devised for the purpose of giving a uniform rate of speed to the movement of wheels.
Essential.	The important feature; the principal thing.
Expansion.	To enlarge; growing greater.
Equidistant.	The same distance from a certain point.
Evolved.	Brought out of; the result of certain considerations.
Facet.	A face.
Facilitated.	Made easy.
Flux.	Any substance which will aid in uniting material under heat. The act of uniting.
Fluid.	Any substance in which the particles freely interchange positions.
Flour emery.	Emery which is finely ground.
Flexible.	The quality of any material which will permit bending.
Float cut.	The term when applied to a tool where the cut is an easy one.
Flexure.	The springing yield in a substance.
Foot pound.	A unit, usually determined by the number of pounds raised one foot in one second of time. 550

198

	pounds raised one foot in one second of time, means so many foot pounds. p. 202
Formulate.	To arrange; to put in order from a certain consideration of things.
Focus.	The center of a circle.
Foci.	One of the points of an ellipse.
Formation.	The structure of a machine or of a compound.
Fractured.	Broken.
Fundamental.	Basis; the first form; the original structure.
Fulcrum.	The resting place for a lever.
Fusion.	Melting. The change of a metal from a solid to a liquid state by heat.
Fusible.	That which is capable of being melted.
Fulminate.	A substance that will ignite or explode by heat or friction.
Gamut.	The scale of sound or light, or vibrations of any kind.
Gear.	A toothed wheel of any kind.
Gelatine.	A tasteless transparent substance obtained from animal tissues.
Globular.	Having the form of a globe or ball.
Glazed.	Having a glossary appearance.
Graphite.	A metallic, iron-black variety of carbon.
Graduated.	To arrange in steps; a regular order or series.
Grinder.	Any mechanism which abrades or wears down a substance.
Gullet.	The curved notches or grooves between projecting parts of mechanism.
Harmonizing.	To make the various parts act together in unison.
H. P.	The symbol for horse power.
Helico.	A form resembling that of the threads of a screw.
Hexagon.	Six-sided.

Heliograph.	The system of signaling by using flashlights.
Horizontal.	Things level with the surface of the earth; like the surface of water.
Hydrogen.	The lightest of all the elements. A tasteless, colorless substance. p. 203
Import.	To bear, or convey as a meaning.
Impulse.	The application of an impelling force.
Impact.	A collision; striking against.
Invariably.	Constant; without failing.
Inertia.	The quality of all materials to remain at rest, or to continue in motion, unless acted on by some external force.
Intersect.	To divide at a certain point. The crossing point of one line over another.
Interval.	A space; a distance between.
Intensity.	Strained or exerted to a high degree.
Interstices.	The spaces between the molecules or atoms in a substance.
Intermediate.	Between.
Intermeshing.	The locking together of gear wheels.
Internal.	That which is within.
Inability.	Unable to perform or do.
Initial.	The first; at the start.
Increment.	One of the parts which go to make up the whole.
Inference.	Drawing a conclusion from a certain state of things.
Insoluble.	A substance which cannot be liquefied by a liquid.
Indentations.	Recesses, or cut-out parts or places.
Induction.	The movement of electricity through the air from one conductor to another.
Inflammable.	That which will burn.
Inclining.	At an angle; sloping.

Inconsequential.	Not of much importance.
Isometric.	That view of a figure which will give the relation of all the parts in their proper proportions.
Jaw.	The grasping part of a vise, or other tool.
Joule.	The practical unit of electrical energy.
Key-way.	A groove in a shaft and in the hub of a wheel, to receive therein a locking key. p. 204
Kilowatt.	A unit of electrical power; one thousand watts.
Kinetic.	Consisting of motion.
Lacing.	The attaching of the ends of a belt to each other.
Lap.	A tool, usually of copper or lead, on which flour emery is spread, with oil, and used to grind out the interior of cylinders.
Lapping.	The act of using a lap to grind out cylinders.
Lacquer.	A varnish for either wood or metal.
Lazy-tongs.	A form of tool, by means of which a long range of movement is attainable, and great grasp of power.
Levigated.	Reduced to a fine powder.
Litharge.	A form of lead used in paints for drying purposes.
Longitudinal.	Lengthwise.
Luminous.	That which has the capacity to light up.
Magnet.	A bar of iron or steel that has electricity in it capable of attracting certain metals.
Manipulation.	Capable of being handled.
Mandrel.	The revolving part of a lathe; a rod or bar which turns and carries mechanical elements thereon.
Manually.	Operated by hand.
Margin.	An edge.
Malleability.	Softness. The state of being formed by hammering.
Magnetism.	A quality of certain metals to receive and hold a

charge of electricity.

Major axis.	The measurement across the longest part of an ellipse.
Minor axis.	The distance across the narrowest part of an ellipse.
Meridian.	The time when the sun crosses the middle of the heavens; midday.
Metric.	Measure; a system which takes the unit of its measurement from the circumference of the earth.
Micrometer.	A tool for measuring small spaces or intervals.
Milling machine.	A large tool for the purpose of cutting gears and grooves or surfaces.
Miter.	A meeting surface between two right-angled pieces. p. 205
Momentum.	That quality of matter which is the combined energy of mass and speed.
Molecular.	Any substance that is made up of any particles; the component elements in any substance.
Modifications.	Changes; improved arrangements.
Multiplicity.	Many; numerous; a large quantity.
Mutilated.	As applied to a gear, one in which certain teeth are removed.
Nautical.	Marine; applied to shipping, and the like.
Neutralizes.	Any substance, like a chemical, which, when added to another chemical, will change them both.
Nitro-glycerine.	An explosive made from glycerine and nitrogen.
Oblique.	At an angle; inclined.
Obliterate.	To wipe out.
Obvious.	That which can be seen; easily observed.
Obtuse.	A blunt angle; not noticeable.
Odophone.	An instrument for determining and testing odors.
Olfactory.	The nerves of the sense of smell.
Orifice.	An opening; a hole.

Oscillation.	A movement to and fro, like a pendulum.
Oxygen.	The most universal gas, colorless and tasteless; is called the acid-maker of the universe and unites with all known substances, producing an acid, an alkali, or a neutral compound.
Oxidizing.	To impart to any substance the elements of oxygen.
Oxide.	Any substance which has oxygen added to it.
Pallet.	A part of a tooth or finger which acts on the teeth of a wheel.
Parallel.	Lines or sides at equal distance from each other from end to end. p. 206
Paraffine.	A light-colored substance, produced from refined petroleum.
Perimeter.	The outer margin of a wheel; the bounding line of any figure of two dimensions.
Periphery.	The outer side of a wheel.
Peen.	The nailing end of a hammer.
Persistence.	That quality of all matter to continue on in its present condition.
Perpendicular.	A line drawn at right angles to another.
Perpetual.	Without end.
Perspective.	A view of an object which takes in all parts at one side.
Physically.	Pertaining to the body.
Phonautograph.	An apparatus for recording sound.
Phonograph.	An apparatus for taking and sending forth sound vibration.
Phenomena.	Any occurrence in nature out of the ordinary.
Pitman.	The rod or bar which connects the piston and crank.
Pivot.	A point or bar on which anything turns.
Pinion.	A small toothed wheel.

Pitch.	The number of vibrations. The term used to give the number of teeth in a wheel.
Pitch diameter.	The point from which the measurements are made in determining the pitch.
Pivoted.	A bar, lever, or other mechanical element, arranged to turn on or about a point.
Plastic.	A substance in such a state that it may be kneaded or worked.
Planer.	A large tool designed to cut or face off wood or metal.
Porosity.	The quality in all substances to have interstices, or points of separation, between the molecules.
Potential.	The power.
Properties.	The qualities possessed by all elements.
Projecting.	The throwing forward. The sending out. p. 207
Promulgated.	Put forth; enunciated.
Protractor.	A mechanic's and draughtsman's tool by means of which angles may be formed.
Promote.	To carry forward in a systematic way.
Precision.	Work done with care; observing correct measurements.
Prony brake.	A machine for determining horse power.
Punch.	A small tool to be struck by a hammer in order to make an impression or indentation.
Quadrant.	One-fourth of a circle.
Quadrant plate.	A plate on which are placed lines and numbers indicating degrees.
Quadruplex.	A term to designate that system of telegraphy in which four messages are sent over a single wire at the same time.
Ratchet.	A wheel having teeth at certain intervals to catch the end of a pawl or finger.
Ratchet brace.	A tool to hold a drill, having a reversible ratchet wheel.

Rasp cut.	A cut of a file which is rough, not smooth.
Rake.	The angle or inclination of the upper surface of the cutting tool of a lathe.
Reverse.	To turn about; in the opposite direction.
Reciprocating.	To go back and forth.
Revolve.	To move in an orbit or circle, as a merry-go-round.
Reciprocity.	To give back in like measure.
Reflection.	The throwing back from a surface.
Resonance.	The quality of vibration which adds to the original movement, and aids in perpetuating the sound.
Refraction.	The quality of light which causes it to bend in passing through different substances.
Reducing.	Bringing it down to a smaller compass.
Rectilinear.	A straight line. p. 208
Retort.	A furnace of refractory material to take high heat.
Reamer.	A tool designed to enlarge or to smooth out holes.
Regulation.	To do things in an orderly way; a system which sets forth certain requirements.
Refractory.	Difficult to work, and not easily fused.
Recess.	A hole, or a depression.
Rocking.	A lever which rotates only part way and then moves in the opposite direction.
Rotate.	A spindle which turns round. Compare revolve.
Rosin.	Certain gums; particularly the sap of pine trees.
Roughing.	The taking off of the first coating with a tool.
Saturated.	A soluble substance which cannot be further dissolved by a liquid.
Scribe.	To mark with a tool.
Screw plate.	A tool which has within it means for adjusting different cutting tools.
Section lining.	The marks made diagonally across drawings to

	indicate that the part is cut away.
Shaper.	A large tool for surfacing off material, cutting grooves, and the like.
Shrinkage.	The term applied to metals when cast, as all will be smaller when cold than when cast in the mold.
Slide rest.	The part of the lathe which holds the tool post.
Sonorous.	Having the quality of vibration.
Slotted.	Grooved, or channeled.
Solvent.	That which can be changed from a solid by liquids.
Spelter.	A combination of zinc and copper. A hard solder.
Soldering.	Uniting of two substances by a third, with heat.
Spindle.	A small shaft.
Spur.	The larger of two intermeshing gears.
Socket.	A depression or hole.
Sprocket.	Teeth in a wheel to receive a chain. p. 209
Spiral.	A form wound like the threads of a screw.
Surface plate.	A true surface made of metal, used as a means of determining evenness of the article made.
Sulphate.	Any substance which is modified by sulphuric acid.
Substitute.	An element or substance used for another.
Superposed.	One placed above the other.
Swage.	Tool for the purpose of changing the form in a material.
Swivel.	A point on which another turns.
Surfacing.	Taking off the outer coating or covering.
Tap.	A small drill.
Tapering.	An object with the sides out of parallel.
Tangential.	A line from the periphery of a circle which projects out at an angle.
Tension.	The exertion of a force.

Tenacity.	The property of a material to hang together.
Tempering.	Putting metal in such condition that it will be not only hard but tough as well.
Technical.	Pertaining to the strict forms and terms of an art.
Texture.	That of which the element or substance is composed.
Threads.	The ridges, spiral in form, which run around a bolt.
Theoretically.	The speculative form or belief in a subject.
Tinned.	The term applied to the coating on a soldering iron with a fluxed metal.
Tines.	Small blades. p. 210
Torsion.	The force exerted around an object, like the action of a crank on a shaft.
Tommy.	A lever to be inserted in a hole in a screw head for turning a screw.
Transmitting.	Sending forth; to forward.
Trammel.	A tool for the purpose of drawing ellipses.
Traction.	Drawing; pulling power.
Tripping.	A motion applied to a finger, which holds a pivoted arm, whereby the latter may be swung from its locked position.
Triangular.	Having three sides and three angles.
Transverse.	Across; at right angles to the long direction.
Undercut.	A wall of a groove or recess which is sloping.
Undulatory.	A wave-like motion, applied generally to light and electricity.
Unit.	A base for calculating from.
Unison.	Acting together; as one.
Unsized.	Generally applied to the natural condition of paper or fabric which has no glue or other fixing substance on it.
Vaporising.	To change from a liquid or solid to a gas.

Variation.	Changing into different conditions; unlike forms.
Verge.	The edge; usually applied to the shoulder of a watch spindle, particularly to the escapement.
Vertical.	Up and down. The direction of a plumb line.
Velocity.	The speed of an article through space.
Vitascope.	An instrument for determining the rate of vibration of different substances.
Vibration.	The movement to and fro of all elements, and by means of which we are made sensitive of the different forces.
Vocation.	The business or the calling of a person. p. 211
Warding.	The act of cutting a projection or guard, such as is usually found on the insides of locks, and the correspondent detent in the key.
Watt.	In electricity the unit of the rate of working in a circuit. It is the electro-motive force of one volt and the current intensity of one ampere.